Withdrawn
University of Waterloo

Electromagnetic Biointeraction

Mechanisms, Safety Standards,
Protection Guides

Electromagnetic Biointeraction

Mechanisms, Safety Standards, Protection Guides

Edited by
Giorgio Franceschetti
CNR-IRECE
Naples, Italy

Om P. Gandhi
University of Utah
Salt Lake City, Utah

and

Martino Grandolfo
Istituto Superiore di Sanità
Rome, Italy

Plenum Press • New York and London

Proceedings of an International Course on Worldwide Nonionizing
Radiation Safety Standards: Their Rationale and Problems,
held May 2-6, 1988, on the Island of Capri, Italy

ISBN 0-306-43328-1

© 1989 Plenum Press, New York
A Division of Plenum Publishing Corporation
233 Spring Street, New York, N.Y. 10013

All rights reserved

No part of this book may be reproduced, stored in a retrieval system, or transmitted
in any form or by any means, electronic, mechanical, photocopying, microfilming,
recording, or otherwise, without written permission from the Publisher

Printed in the United States of America

FOREWORD

This book collects the revised lectures held at Capri (Italy) in the period 2-6 May, 1988 in occasion of the International Course on "Worldwide Nonionizing Radiation Safety Standards: Their Rationales and Problems". The Course was organized by IRECE (Institute for Research in Electromagnetism and Electronic Components) of CNR (Italian National Council for Research) and was directed by professors Giorgio Franceschetti and Om P. Gandhi.

The idea for this course arose from the continuing wide disparity in the electromagnetic (EM) radiation safety standards worldwide, and the confusion that this has caused in the public mind. The safety guidelines in the western countries have been nearly three orders of magnitude greater than the safety levels in the Eastern European countries. Even though the former have been slightly reduced and the latter have been increased somewhat in recent years, there is still a wide gap in the EM safety standards that are used. With the ever increasing use of EM energy the public is becoming increasingly aware of and concerned about the potential biohazards of EM fields. This problem is compounded by inadequate knowledge of nonthermal mechanisms of interaction of EM fields with biological systems.

The lecturers for the Course were the recognized leaders in their respective areas within the discipline of Biological Effects of Electromagnetic Fields. The attendees came from a variety of backgrounds and from various geographical locations including Finland, Italy, Poland, Saudi Arabia, Spain, Sweden, The Netherlands, United Kingdom, and the United States. They participated actively in the lively discussions after each of the lectures. The Course was held at one of the most scenic spots in the world - the island of Capri - during the spring season. The abundance and the fragrance of the budding flowers provided a memorable setting for the course.

This book collects the lectures held at Capri after a careful revision, expansion and mutual blending, in order to provide an exhaustive information on the book topic.

The first chapter, "Basic Definitions and Concepts" by Professor Guglielmo d'Ambrosio provides the definitions of physical quantities used in the subsequent part of the book. The following chapters, from the second to the fifth, are relative to the description of the interaction mechanisms between fields and bodies: chapter two, "Advances in RF Dosimetry: Their Past and Projected Impact on the Safety Standards" by Professor Om P. Gandhi is relative to the macroscopic approach; chapter three, "Interaction Mechanisms at Microscopic Level" by Professor Paolo Bernardi and Professor Guglielmo D'Inzeo, and four, "Biological Effects of Radiofrequency Radiation: An Overview" by Professor Stephen F. Cleary, consider the microscopic point of view; chapter five ,"Electromagnetic

Fields and Neoplasms" by Professor Stanislaw Szmigielski and Dr. Jerzy Gil is mainly related to the epidemiological point of view. The subseguent chapters, from the sixth to the eight, represent the core of the book, discussing the existing safety standards and protection guides: in chapter six, "Worldwide Public and Occupational Radiofrequency and Microwave Protection Guides" by Dr. Martino Grandolfo and Dr. Kjell Hansson Mild a comparative analysis is presented of the existing or recently proposed standards; chapter seven, "The Rationale for the Eastern European Radiofrequency and Microwave Protection Guides" by Professor Stanislaw Szmigielki and Dr. Tadeusz Obara is relative to the rationale underlyng stamdards issued in the Eastern European countries; chapter eight, "Existing Safety Standards for High-Voltage Transmission Lines" by Dr. Martino Grandolfo and Dr.Paolo Vecchia, considers the standards relative to 50/60 Hz power line fields. The last two chapters are concerned about the numerical and experimental determination of the fields. The experimental approach is considered in chapter nine, "Instumentation for Electomagnetic Fields Exposure Evaluation and Its Accuracy" by Dr. Santi Tofani and Dr. Motohisa Kanda; the theoretical approach is considered in chapter ten, "Numerical Methods" by Professor Om P. Gandhi.

We believe that this book is useful for all the bioelectromagnetic scientific community as well as to scientific and technical people concerned with business which require application, monitoring and test of safety standards. We hope that the book will meet the same success of the Course; and we are indebted to many people and Institutions that contributed to both the Course organization and the book realization.

As far as the Course is concerned, we would like to acknowledge the excellent local arrangements made by Mr. Generoso Sole and dr. Maria Rosaria Scarfi'. Their attention to minute details and their painstaking labor contributed much to the smooth functioning of the course. We would also like to thank the speakers for their excellent and provocative lectures and for the timely submissions of their manuscripts first in the draft form and finally in the camera-ready format needed to this book. We would also like to extend our thanks to the sponsoring agencies that contribute finantially to the budget of the Course: first of all the Italian CNR through its research institute IRECE; and then the INFN, Sezione Sanita', Roma; Regione Campania, Assessorato alla Sanita' and Servizio Istruzione e Cultura; University of Naples (central offices); Department of Electronic Engineering, University of Naples. Other Institutions supported the Course, and we extend our thanks to them also: International Nonionizing Radiation Comitee of International Radiation Protection Association; Istituto Superiore di Sanita'; University of Utah.

Coming to the book, the dedication of all the Authors should be commended: we thank all of them. We also like to thank Ms. Janie Curtis of Plenum Press for her continuous assistance and Ms. Consiglia Rasulo and again Mr. Generoso Sole for their editorial help. The excellent secretarial assistance of Ms. Alma Paoluzzi and the help of Ms. Annamaria Lopomo in revising English style of some chapters are gratefully acknowledged, with our deepest gratitude for their long hours of work well above the normal call of duty.

            Giorgio Franceschetti (Napoli, Italy)

            Om P. Gandhi (Salt Lake City, Usa)

            Martino Grandolfo (Roma, Italy)

March 31, 1989

CONTENTS

Some Basic Definitions and Concepts 1
   G. d'Ambrosio

Advances in RF Dosimetry: Their Past and Projected
   Impact on the Safety Standards 11
   O.P. Gandhi

Interaction Mechanisms at Microscopic Level 27
   P. Bernardi, and G. D'Inzeo

Biological Effects of Radiofrequency 59
   S.F. Cleary

Electromagnetic Fields and Neoplasms 81
   S. Szmigielski, and J. Gil

Worldwide Public and Occupational Radiofrequency
   and Microwave Protection Guides 99
   M. Grandolfo, and K. H. Mild

The Rationale for the Eastern European Radiofrequency
   and Microwave Protection Guides 135
   S. Szmigielski, and T. Obara

Existing Safety Standards for High Voltage
   Transmission Lines 153
   M. Grandolfo, and P. Vecchia

Instrumentation for Electromagnetic Fields
   Exposure Evaluation and Its Accuracy 175
   S. Tofani, and M. Kanda

Numerical Methods 193
   O.P. Gandhi

Contributors 215

Index 217

SOME BASIC DEFINITIONS AND CONCEPTS

G. d'Ambrosio

Dept. of Electronic Engineering
University of Naples
Naples, Italy

BASIC ELECTROMAGNETIC QUANTITIES

The force F, acting on an electric charge q moving with velocity v depends on two vector quantities: the *electric field E*, and the *magnetic induction B*, according to the following relationship:

$$F = q(E + v \times B) \ . \tag{1}$$

All above quantities are depending on position P and time t. Here a general time dependence is assumed and italic letter symbols are used. In the following section time-harmonic fields will be considered and roman letter symbols will be introduced. Once the force F, the charge q, and the velocity v are assumed to be measurable quantities, vectors E and B can be defined by means of the above formula (for instance E coincides with the force acting on the unit positive charge at rest ($v=0$)) (Feynman et al., 1963). The unit of force and charge being respectively one newton [N] and one coulomb [C], the unit of electric field E is one newton per coulomb. The same unit is also one volt per meter [V/m]. The unit of magnetic induction B is one (volt·second) per meter^2. It is also called one tesla [T].

The electric field, E, and the magnetic induction, B, obey the following relationship, written in its differential and integral forms:

$$\nabla \times E = -\frac{\partial B}{\partial t} \ , \tag{2}$$

$$\oint_C E \cdot d\mathbf{C} = -\frac{\partial}{\partial t} \iint_S B \cdot d\mathbf{S} \ , \tag{2'}$$

where $\partial/\partial t$ is the time partial derivative, and C is the contour of the fixed surface S; vector $d\mathbf{C}$ is tangent to C, and vector $d\mathbf{S}$ is normal to S: they are directed as shown in fig.1.

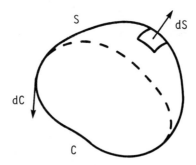

Fig.1. A fixed surface S and its contour C. Vector $d\mathbf{S}$, normal to S, and vector $d\mathbf{C}$, tangent to C, are also shown with proper associate directions.

Strictly related to the above electric field $\mathbf{E}$ and magnetic induction $\mathbf{B}$ are two other vector quantities, namely the electric induction $\mathbf{D}$ and the magnetic field $\mathbf{H}$, that obey the following relationship, written in its differential and integral forms:

$$\nabla \times \mathbf{H} = \frac{\partial \mathbf{D}}{\partial t} + \mathbf{J} , \tag{3}$$

$$\oint_C \mathbf{H} \cdot d\mathbf{C} = \frac{\partial}{\partial t} \iint_S \mathbf{D} \cdot d\mathbf{S} + \iint_S \mathbf{J} \cdot d\mathbf{S} , \tag{3'}$$

where $\mathbf{J}$ is the current density (unit: one ampere per meter^2 [A/m^2]). The unit of electric induction $\mathbf{D}$ is one (ampere-second) per meter^2 = one coulomb per meter^2 [C/m^2], and the unit of magnetic field $\mathbf{H}$, is one ampere per meter [A/m]. In time invariant, linear, isotropic, non dispersive media the above $\mathbf{E}$, $\mathbf{B}$, $\mathbf{H}$, $\mathbf{D}$, $\mathbf{J}$ (functions of position and time) obey also the following relationships:

$$\begin{aligned} \mathbf{J} &= \sigma \mathbf{E} , \\ \mathbf{D} &= \epsilon \mathbf{E} , \\ \mathbf{B} &= \mu \mathbf{H} , \end{aligned} \tag{4}$$

where σ, ϵ, and μ are scalar quantities depending on position. σ is the conductivity; the unit of conductivity is one ampere per (meter·volt) = one (ohm·meter)$^{-1}$. It is also called one siemens per meter [S/m]. ϵ is the permittivity; the unit of permittivity is one coulomb per (meter·volt) = one farad per meter [F/m]. In free space $\epsilon = \epsilon_o = 8.854 \times 10^{-12}$ F/m. μ is the permeability; the unit of permeability is one (tesla·meter) per ampere = one (ohm·second) per meter. It is also called one henry per meter [H/m]. In free space and in all non-magnetic materials, included the biological ones, $\mu = \mu_o = 4\pi \times 10^{-7}$ H/m $\simeq 1.3$ microhenry per meter, so that a magnetic field $|\mathbf{H}| = 1$ A/m corresponds to a magnetic induction $\mu_o|\mathbf{H}| = |\mathbf{B}| \simeq 1.3$ microtesla.

TIME-HARMONIC FIELDS

Let all the scalar components of the above vector quantities be sinusoidal functions of time. For instance:

$$E_x(P,t) = E_{xM}(P)\cos[\phi_x(P) + \omega t] ,\tag{5}$$

where the amplitude, E_{xM}, and the phase at time $t = 0$, ϕ_x, are functions of the point P. ω ($= 2\pi/T = 2\pi f$) is the angular frequency, T is the time period and f is the frequency. The unit of time being one second [s], the unit of frequency is one cycle per second [s^{-1}]. It is also called one hertz [Hz]. The *root mean square (rms)* value is given by:

$$E_{x\;rms}(P) = \sqrt{\frac{1}{T}\int_0^T E_x^2(P,t)dt} = \frac{1}{\sqrt{2}} E_{xM}(P) .\tag{6}$$

Equation (5) can be rewritten as:

$$E_x(P,t) = \sqrt{2} E_{x\;rms}(P)\cos[\phi_x(P) + \omega t] \tag{7}$$

or:

$$E_x(P,t) = \sqrt{2}\mathrm{Re}\left\{E_{x\;rms}(P)\exp[j\phi_x(P)]\exp(j\omega t)\right\} ,\tag{8}$$

where Re means "real part of". By putting $E_{x\;rms}(P)\exp[j\phi_x(P)] = \mathrm{E_x}(P)$, equation (8) can be rewritten as:

$$E_x(P,t) = Re[\mathrm{E_x}(P)\exp(j\omega t)] .\tag{9}$$

In vector form one has:

$$\boldsymbol{E}(P,t) = \sqrt{2}Re[\mathbf{E}(P)\exp(j\omega t)] ,\tag{10}$$

where the complex vector $\mathbf{E}(P)$ is the *rms phasor* of the electric field.

In time invariant, linear, isotropic, possibly dispersive media

$$\begin{aligned}\mathbf{J} &= \sigma(\omega)\mathbf{E} ,\\ \mathbf{D} &= \epsilon(\omega)\mathbf{E} ,\\ \mathbf{B} &= \mu(\omega)\mathbf{H} ,\end{aligned}\tag{11}$$

where roman letters indicate phasor quantities. In terms of such phasor quantities, equation (3) becomes:

$$\nabla \times \mathbf{H} = j\omega\epsilon\mathbf{E} + \sigma\mathbf{E} = j\omega(\epsilon + \frac{\sigma}{j\omega})\mathbf{E} .\tag{12}$$

Let it be (Collin, 1960; Harrington, 1961):

$$\epsilon = \epsilon_1 - j\epsilon_2 ,$$
$$\epsilon_1 = \epsilon_o \epsilon' ,$$
$$\epsilon_2 + \sigma/\omega = \epsilon_o \epsilon'' = \sigma_{eq}/\omega .$$
(13)

The above eq.(12) can be rewritten as:

$$\nabla \times \mathbf{H} = j\omega\epsilon_o(\epsilon' - j\epsilon'')\mathbf{E} ,$$

where $(\epsilon' - j\epsilon'')$ is a complex equivalent relative permittivity. σ takes into account ohmic losses caused by migrating charge carriers, ϵ_2 takes into account dielectric losses due to the motion of bound charges, $\sigma_{eq}(=\omega\epsilon_o\epsilon'')$ takes into account both kinds of losses; the power dissipated per unit volume is given by:

$$\sigma_{eq}|\mathbf{E}|^2 = \omega\epsilon_o\epsilon''|\mathbf{E}|^2 \quad \text{watt per meter}^3 \ [\text{W/m}^3]. \tag{14}$$

The equivalent conductivity, σ_{eq}, is sometimes indicated simply as σ (von Hippel, 1954; Stuchly, 1983; Foster and Schwan, 1986), although it is not a pure conductivity.

Time-harmonic electromagnetic fields find different uses depending on the frequency range. In fig.2 the electromagnetic spectrum is shown together with some of the most important applications.

TIME-HARMONIC PLANE WAVES

A time-harmonic homogeneous plane wave is a time-harmonic electromagnetic field that varies in space only along a given direction (say z). This space distribution has a sinusoidal shape that travels (propagates) along that direction. The vectors $\mathbf{E}$ and $\mathbf{H}$ are mutually perpendicular and perpendicular to the direction of propagation.

A time-harmonic plane wave is said to be linearly polarized when the directions of the field vectors are always and everywhere the same. Along the direction of propagation (z) the phase changes linearly with space $[\exp(-j\beta z)]$ according to a phase constant $\beta = 2\pi/\lambda$. The so called wavelength, λ, is the space period of the phase, and $\lambda/T = \lambda f$ is the phase velocity. The unit of β is one radian per meter [rad·m^{-1}]. Along the direction of propagation the amplitude may change too, according to $\exp(-\alpha z)$, α being the attenuation constant. The unit of α is one neper per meter [m^{-1}]. The reciprocal of $\alpha(1/\alpha = \delta)$ is the distance over which the field reduces by a factor $1/e \simeq 0.37$ (~ 8.7 dB). The above phase and amplitude constants, and the complex propagation constant $k = \beta - j\alpha$ depend on the parameters of the medium; in biological media ($\mu = \mu_o$):

$$k = \beta - j\alpha = \omega\sqrt{(\epsilon' - j\epsilon'')\epsilon_o\mu_o} = k_o\sqrt{\epsilon' - j\epsilon''} , \tag{15}$$

$$\beta = k_o\sqrt{\frac{|\epsilon_r| + \epsilon'}{2}} , \tag{16}$$

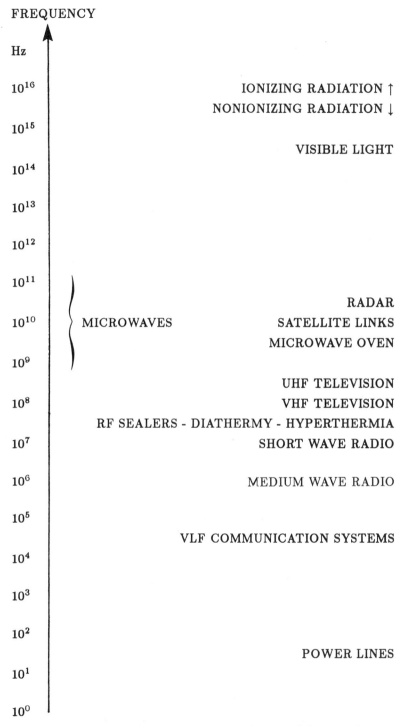

Fig.2. The electromagnetic spectrum and some of the most important applications.

$$\alpha = k_o \sqrt{\frac{|\epsilon_r| - \epsilon'}{2}} ,\tag{17}$$

where $k_o = \omega\sqrt{\epsilon_o\mu_o}$ (free space propagation constant), and $|\epsilon_r| = \sqrt{\epsilon'^2 + \epsilon''^2}$.

In plane waves a power flux occurs along the direction of propagation, and over each plane perpendicular to this direction the complex power surface density, in case of linear polarization, is given by:

$$|\mathbf{E}|^2/\varsigma = \varsigma|\mathbf{H}|^2 \quad [\text{W/m}^2] ,\tag{18}$$

where

$$\varsigma = \sqrt{\frac{\mu}{\epsilon}}$$

is named the *intrinsic impedance of the medium*. The unit of ς is one ohm [Ω]. The intrinsic impedance of the free space is:

$$\varsigma_o = \sqrt{\frac{\mu_o}{\epsilon_o}} = 377 \ \Omega .\tag{19}$$

In biological media:

$$\varsigma = \sqrt{\frac{\mu_o}{\epsilon_o(\epsilon' - j\epsilon'')}} .\tag{20}$$

In the case $\epsilon' \ll \epsilon''$ (high loss tissues):

$$\varsigma \simeq \sqrt{\frac{\mu_o}{-j\epsilon_o\epsilon''}} = \sqrt{\frac{j\omega\mu_o}{\sigma_{eq}}} .\tag{21}$$

In case of a discontinuity (two half-spaces, joined along a plane and having intrinsic impedances ς_1 and ς_2), instead of a single plane wave one must consider, in general, an incident wave, a reflected wave and a transmitted wave. In case that the discontinuity plane is perpendicular to the direction of propagation (normal incidence) the reflected electric field, $\mathbf{E}_r$, can be easily calculated from the incident one, $\mathbf{E}_i$, via the reflection coefficient Γ:

$$\mathbf{E}_r = \Gamma\mathbf{E}_i ,\tag{22}$$

$$\Gamma = \frac{\varsigma_2 - \varsigma_1}{\varsigma_2 + \varsigma_1} .\tag{23}$$

The real part of the incident power density times $|\Gamma|^2$ gives the real part of the reflected power density, and the real part of the incident power density times $(1 - |\Gamma|^2)$ gives the real part of the transmitted power density.

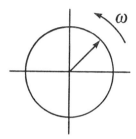

Fig.3. Counter-clockwise circular polarization. The propagation vector points toward the observer.

Let us have two plane waves at the same frequency, propagating along the same direction, and linearly polarized along two directions (let us think, for instance to the **E** field directions) mutually perpendicular. If there is a phase difference $\Delta\phi = n\pi$ ($n = 0, 1, 2, ...$), by superposition one gets a new wave, linearly polarized along a direction depending on the amplitudes of the two component waves. If, on the contrary, there is a phase difference $\Delta\phi = \pi/2 + n\pi$ and the amplitudes of the two waves are the same, the resulting field has constant amplitude, but its direction changes according to the phase $(\omega t - \beta z)$ (for instance at a given point z, the field rotates with angular velocity ω: circular polarization, see fig.3). In case the amplitudes of the two waves are not the same, or the phase difference is neither $n\pi$ nor $\pi/2 + n\pi$, the resulting field is ellyptically polarized, i.e. at a given point, z, it rotates by varying periodically both amplitude and angular velocity.

THE "FAR FIELD" REGION

In the region of space very close to an antenna, or, more generally, to a radiofrequency or microwave source, the electric and magnetic vectors vary according to complex patterns. It is mainly in this region, bounded up to a distance less than, say λ, that the reactive phenomena take place (*reactive field region*).

The space outside the above region is characterized by fields that contribute to the power radiated by the antenna (*radiative field region*). Also in this region the field amplitude may still oscillate if one moves along a prescribed direction away from the source; this is due to the change in the phase difference, between the contributions arising from the various antenna surface elements. Here we are in the *radiative near field region*. The *radiative far field region* starts at a distance such that the above changes in the phase differences become negligible, and along a given direction the field amplitude decays monotonically (Hansen, 1964).

At distances r, along the axis of the antenna (fig.4), much more larger than the dimension D ($r >> D$), the further condition $r > 2D^2/\lambda$ ensures that the maximum path difference Δr (fig.4) is less than $\lambda/16$. In the microwave range the antenna is usually many wavelengths wide ($D >> \lambda$); in this case $2D^2/\lambda$ is $>> D$ ($>> \lambda$) and an increase in distance r beyond the $2D^2/\lambda$ limit has, as a consequence, a field amplitude decrease like $1/r$ (Hansen, 1964). In the microwave range the above distance ($r =$

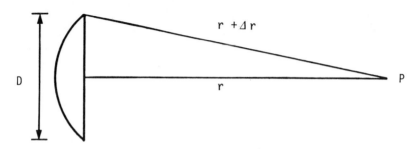

Fig.4. A circular aperture transmitting antenna and a receiving point P at distance r on the axis of the antenna. D is the antenna diameter and Δr is the maximum path difference between the various aperture elements and the receiving point.

$2D^2/\lambda$) is usually assumed as the bound of the far field region. At lower frequencies (at MF, for instance) the antenna may be short in terms of wavelength ($D << \lambda$). In this case $r = 2D^2/\lambda$ ($<< D << \lambda$) is a distance falling within the reactive region, and a larger distance must be chosen as the far field region bound. A suitable choice in the above conditions can be $r = \lambda$, and a general rule, for all frequencies, is to assume the largest distance between $2D^2/\lambda$ and λ as the border between near- and far-field regions.

EXPOSURE AND ABSORPTION PARAMETERS

The electromagnetic field existing in a given region of the free space is said the *incident field* with reference to a body that should subsequently be placed in that region. It has to be stressed that, once the body is actually put in that region the actual field both inside and outside the body may be very different from the incident one. The incident field describes the *exposure conditions* of a body. By definition these exposure conditions are depending on the source and on the position with respect to the source, but are not depending on the exposed body.

At distances r from a radiating antenna much more larger than both the wavelength ($r >> \lambda$) and the antenna width ($r >> D$, fig.4) the antenna may be viewed as a point-like source S radiating a spherical wave. At these distances the illumination over a transverse dimension d (fig.5) that is small with respect to the distance ($d << r$) and satisfies the further condition $d < \sqrt{r\lambda/2}$ ($r > 2d^2/\lambda$), is nearly equiphase. In fact the maximum path difference Δr over d is less than $\lambda/16$ and the maximum phase difference is less than $(2\pi/\lambda)(\lambda/16) = \pi/8$. In addition, for most practical sources, in the above conditions the amplitude taper over d is negligible, so that a nearly plane wave illumination over d is obtained. In free space the plane wave power density is real and is given by either $|\mathbf{E}|^2/\zeta_o$ or $\zeta_o|\mathbf{H}|^2$ (linear polarization, eq.(18)), so that in the above far-field plane-wave conditions the strengths of both the incident electric and magnetic fields can be easily calculated once the power density is known. On the contrary in the near zone, where a more complex pattern takes place, electric and magnetic fields must be independently specified at every point.

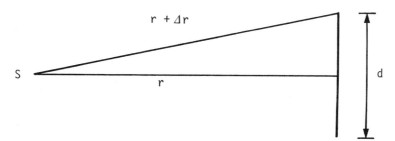

Fig.5. A point-like source S at distance r on the axis of a plane circular target having diameter d. Δr is the maximum path difference between the source and the various elements of the surface of the target.

Once the body is exposed to the field, some power P [W] can be absorbed by the body itself, and this power depends on both the exposure conditions and the exposed body. The absorbed power P, divided by the volume V of the body gives the (whole-body) *average absorbed power volume density* expressed in watt per meter^3 [W/m^3]. With reference to a vanishing small volume element around a given point in the body, we have the ratio dP/dV that gives the *local absorbed power volume density*.

As said before - see eq.(14) - in a material medium having no magnetic losses but complex permittivity $\epsilon_o(\epsilon' - j\epsilon'')$ the local volume density of the absorbed power can be related to the local (internal) electric field rms strength $|\mathbf{E}|$ as follows:

$$dP/dV = \sigma_{eq}|\mathbf{E}|^2 = \omega\epsilon_o\epsilon''|\mathbf{E}|^2 \quad [\text{W/m}^3] \ . \tag{24}$$

If one refers the absorbed power P to the body mass M instead of the volume, one has the (whole-body) *average absorbed power mass density* expressed in watt per kilogram [W/kg], and the *local absorbed power mass density*, dP/dM. The (whole-body average, or local) absorbed power mass density is usually indicated as the whole-body average, or local, *specific absorption rate* (SAR). The local SAR at the human body surface may be very high at millimeter wavelengths (Gandhi and Riazi, 1986) and grows linearly with frequency in the high frequency limit (Franceschetti, 1988).

The absorbed power P, we were dealing with, may vary in time due to changes in exposure conditions (incident field) or due to changes (e.g. orientation) of the body with respect to the incident field. The former case may occur due to specific features of the source: for instance radar generators give high power levels but only during short time intervals followed by longer zero power intervals (fig.6). In such pulsed wave conditions we have a peak power value P_M and an average power value P_{AV}:

$$P_{AV} = P_M \cdot \tau/T \ , \tag{25}$$

where τ/T is the pulse duty cycle. More generally, once the time dependence of the absorbed power $P = P(t)$ is known, one obtains the *absorbed energy (dose)* by integration over a time interval (t_1, t_2):

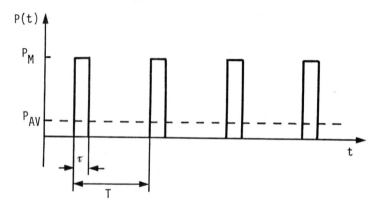

Fig.6. Power pattern of a radar generator. T = period. τ = pulse width. τ/T = duty cycle.

$$\int_{t_1}^{t_2} P(t)dt \quad \text{watt} \cdot \text{second} = \text{joule} \quad [J] \;, \tag{26}$$

and the *average absorbed power*:

$$\frac{1}{t_2 - t_1} \int_{t_1}^{t_2} P(t)dt \quad \text{watt} \; [W] \;. \tag{27}$$

REFERENCES

Collin R.E., 1960, "Field Theory of Guided Waves", McGraw-Hill, New York.

Feynman R.P., Leighton R.B., and Sands M., 1963, "The Feynman Lectures on Physics", Addison-Wesley, London.

Foster K.R., and Schwan H.P., 1986, Dielectric properties of tissues, *in*: "CRC Handbook of Biological Effects of Electromagnetic Fields", C. Polk and E. Postw, Eds., CRC Press, Boca Raton.

Franceschetti G., 1988, Introductory Remarks, *in*, " Worldwide Nonionizing Radiation Safety Standards International Course", Lecture Notes, O.P. Gandhi and G. Franceschetti, Eds., Capri.

Gandhi O.P., and Riazi A., 1986 "Absorption of millimeter waves by human beings and its biological implications", *IEEE Trans. on MTT*, 34:228.

Hansen R.C., 1964, Aperture theory, *in*: "Microwave Scanning Antennas", R.C. Hansen, Ed., Academic Press, New York.

Harrington R.F., 1961, "Time-Harmonic Electromagnetic Fields", McGraw-Hill, New York.

Stuchly M.A., 1983, Fundamentals of the interactions of radio-frequency and microwave energies with matter, *in*: "Biological Effects and Dosimetry of Nonionizing Radiation", M. Grandolfo, S.M. Michelson, and A. Rindi, Eds., Plenum Press, New York.

Von Hippel A.R., 1954, "Dielectrics and Waves", Wiley, New York.

ADVANCES IN RF DOSIMETRY: THEIR PAST AND PROJECTED IMPACT ON THE SAFETY STANDARDS

Om P. Gandhi

Department of Electrical Engineering
University of Utah
Salt Lake City, Utah, 84112, U. S. A.

INTRODUCTION

The expanding usage of electromagnetic (EM) radiation has necessitated an understanding of its interaction with humans. Such knowledge is vital in evaluating and establishing radiation safety standards, determining definitive hazard levels, and understanding several of the biological effects that have been reported in the literature. Studies of the effects of EM radiation have used laboratory animals such as rats, rabbits, etc. for the study of biological and/or behavioral effects. For these experiments to have any projected meanings for humans, it is necessary to be able to quantify the whole-body power absorption and its distribution for the various irradiation conditions. It is further necessary that dosimetric information be known for humans subjected to irradiation at different frequencies and for realistic exposure conditions.

Unlike ionizing radiation, where the absorption cross section of the biological target is directly related to its physical cross section, the whole-body EM energy absorption has been shown to depend strongly on polarization (orientation of electric field **E** of the incident waves), frequency, and physical environments such as the presence of ground and other reflecting surfaces. A couple of excellent review articles (Gandhi, 1980; Durney, 1980) summarize the highlights of this work. Some of the most important findings are:

A. Maximum energy is absorbed for incident electric field along the height of the human body for frequencies such that the height h is approximately 0.36 to 0.4 times the wavelength (λ) for free space irradiation. For a 1.75 m-tall individual this corresponds to a frequency on the order of 65 to 70 MHz. Power absorbed under these conditions is 4.2 times larger than that projected from physical cross-section considerations. This is the resonant region of absorption of EM radiation (Gandhi, 1975).

B. For frequencies below resonance, an f^2-type dependence has been observed for the energy absorbed, where f is the frequency of radiation (Durney et al., 1978). For the post-resonant region, the absorbed power reduces as $1/f$, approaching gradually the value that would correspond to roughly one-half that based on physical cross-sectional considerations.

C. For a human standing on a high conductivity ground, the frequency for maximum absorption is approximately one-half that without the ground; i.e., about 35 MHz. At the new resonant frequency, the power absorbed is somewhat larger and corresponds to roughly 8 times that projected from physical cross-section considerations (Gandhi et al., 1977; Hill, 1984; Guy and Chou, 1982).

The above results have had an impact on the 1982 reformulation of the American National Standards Institute (ANSI) guideline (1982) as well as the guidelines formulated by the International Nonionizing Radiation Committee of the International Radiation Protection Association (IRPA-1984). The ANSI-1982 guideline for safety levels with respect to human exposure to radio-frequency (RF) electromagnetic fields, 300 kHz to 100 GHz, is shown in Fig. 1. The protection guide recommends as safe a power density level of 100 mW/cm^2 (electric field E = 614 V/m), for the frequency region 0.3 to 3.0 MHz, with a safety level reducing as $900/f_{MHz}^2$ mW/cm^2 (E = 1842 f_{MHz} V/m, where f_{MHz} is the frequency in MHz) for the frequency region 3.0 to 30.0 MHz to a valley of power density P = 1 mW/cm^2 (E ~ 61.4 V/m) for the frequency band 30 to 300 MHz. The ANSI guide has been prescribed to ensure that the whole-body-averaged SAR shall not exceed 0.4 W/kg for any of the human sizes and age groups. Dosimetric information on EM energy absorption in human beings (Gandhi, 1975, 1979; Durney et al., 1978) was used to obtain the power density as a function of frequency so that under the worst-case circumstances (E-field along the height of the body; grounded and ungrounded conditions), the whole-body-averaged SAR will be less than 0.4 W/kg. Recognizing the highly nonuniform nature of SAR distribution including some regions where there may be fairly high local SARs, the ANSI guide further prescribes that the local SAR "in any 1-g of tissue" shall not exceed 8 W/kg. A recent report by National Council on Radiation Protection (NCRP-1986) has recommended lower power densities that are one-fifth of the values given in the ANSI guideline (1982) for public exposures.

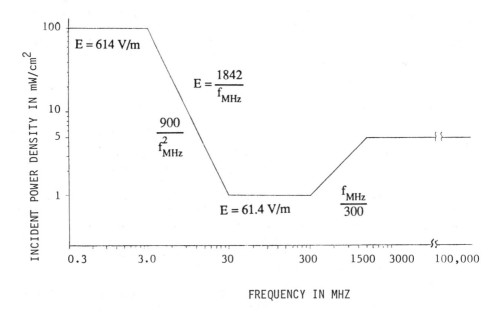

Fig. 1.   ANSI C95.1-1982 safety guideline for human exposure to RF electromagnetic fields.

Recent studies have pointed to several problems with the ANSI RF safety guideline, particularly in the VLF to HF range of frequencies. These are highlighted in the following:

A. The high electric fields sanctioned in the ANSI guideline for LF to HF frequencies will result in significant RF currents flowing through the human body resulting in high SARs in smaller cross-section areas of the body such as the leg and the ankle region (Gandhi et al., 1985a, 1986). Based on measurements with standing human subjects for plane-wave fields, induced currents on the order of 628-780 mA and resulting ankle-section SARs as high as 182-243 W/kg are projected for 1.75 m-tall individuals for the ANSI guideline for the frequency band 3-40 MHz. Using electromagnetic scaling concepts, SARs as high as 371 and 534 W/kg are projected for ten- and five-year-old children, respectively, for f = 50.7 and 62.5 MHz. These values are considerably larger than 8 W/kg for "any 1-g of tissue" assumed in the ANSI guideline and will in fact result in high rates of heating as measured at the surface of the ankle section (Chen and Gandhi, 1988).

B. Commonly encountered ungrounded objects such as car, van, bus, etc. will develop open-circuit voltages on the order of several hundred volts exposed to ANSI-recommended electric field of 614 V/m for the frequency band 0.3 to 3 MHz. Upon touching such vehicles, large currents may flow through the human body that are considerably in excess of those needed for perception, pain, and even burns in some cases (Gandhi and Chatterjee, 1982; Guy and Chou, 1982; Chatterjee et al., 1986). For example, the current flowing through the hand of a human upon holding the door handle of an ungrounded automotive van is estimated to be 879 mA, resulting in a local SAR in the wrist of about 1045 W/kg (Chatterjee et al., 1986).

CURRENTS INDUCED IN A STANDING HUMAN BEING FOR VERTICALLY POLARIZED PLANE-WAVE EXPOSURE CONDITIONS

We have previously shown (Gandhi et al., 1985a, 1986) that the current I_h flowing through the feet of a standing, grounded human being is given by:

$$\frac{I_h}{E} = 0.108 \, h_m^2 \, f_{MHz} \, \frac{mA}{V/m} \tag{1}$$

where E is the plane-wave incident electric field (assumed vertical) in V/m; h_m is the height of the individual in meters; and f_{MHz} is the frequency in MHz. It is interesting to note that the current in Eq. (1) can be considered as though all the fields falling on an area $1.936 \, h_m^2$, or approximately 5.93 m² for a human of height 1.75 m, were effectively passed through the human body. Similar results have also been reported by a number of authors, most notably by Hill and Walsh (1984; to 10 MHz), Tell et al. (to 1.5 MHz), Guy and Chou (1982; at 0.146 MHz), and Gronhaug and Busmundrud (1982; to 27.0 MHz). We have found Eq. (1) to be valid to a frequency of 40 MHz for a 1.75-m human. Because of the f^2 dependence, currents as high as 13.0 mA/(V/m) have been measured at 40 MHz for adult humans (Gandhi et al., 1986). Such high currents have also been recently corroborated by Guy (1987). This finding implies a current, I_h, of nearly 800 mA for the ANSI recommended E-field of 61.4 V/m at this frequency. Anatomical data (Morton et al., 1941) have been used to estimate the effective areas for current flow for the various cross sections of the leg (Gandhi et al., 1985a). The effective area A_e is estimated by the equation:

$$A_e = \frac{A_c \sigma_c + A_\ell \sigma_\ell + A_m \sigma_m}{\sigma_c} \tag{2}$$

where A_c and A_ℓ and A_m are the physical areas of the high-water-content and the low-water-content tissues, and of the region containing red marrow, of conductivities σ_c, σ_ℓ, and σ_m, respectively. Noting that most of the ankle cross section consists of low-conductivity bone or tendon, an effective area of 9.5 cm² is calculated for this cross section for a human adult even though the physical cross section is on the order of 40 cm² (Gandhi et al., 1986). The SARs have been calculated from the equation

$$J^2/\sigma_c\rho = (I_h/2) A_e^2 \sigma_c\rho \tag{3}$$

where ρ is the mass density of the tissue, taken to be 10^3 kg/m^3. The current taken for each of the legs is $I_h/2$. For the E-fields recommended in the ANSI standard (Fig. 1), the ankle-section SARs are calculated by using the current I_h from Eq. (1), and the conductivity values for high-water-content tissues from Johnson and Guy (1972). The SARs thus calculated are shown in Fig. 2. A fairly large ankle-section SAR of 243 W/kg is projected for a standing adult of height h = 1.75 m at a frequency of 40 MHz. This is, of course, considerably higher than the ANSI guideline of 8 W/kg for any 1 g of tissue.

We have previously quantified the currents that would flow through the feet with rubber- or leather-soled shoes (Gandhi et al., 1985a, 1986). The currents for subjects wearing leather-soled shoes and rubber-soled shoes are, respectively, about 90-95 percent and 60-80 percent (the larger values have been measured for higher frequencies) of those for bare feet. The SARs in Fig. 2 must therefore be multiplied by the square of the appropriate current reduction factors.

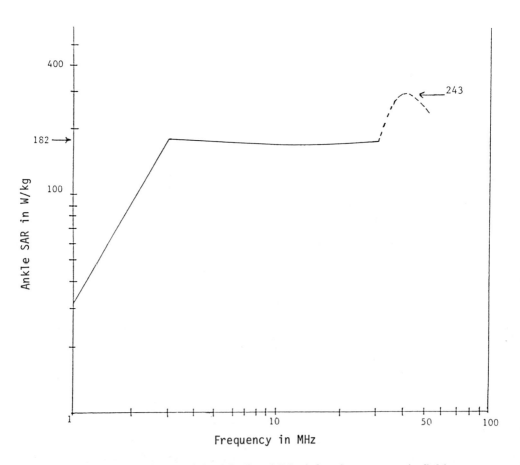

Fig. 2. Ankle SAR for an adult male (h = 1.75 m) for electromagnetic fields recommended in the ANSI-1982 RF safety guideline.

Scaling to Other Heights

Equation (1) demonstrates that the induced current is proportional to h^2, and is consequently smaller for shorter individuals. Since the cross-sectional dimensions of the body are to a first order of approximation also proportional to (weight)$^{2/3}$ or h^2, the current density J and hence the ankle-section SAR ($= J^2/\sigma_c\rho$) may not be very different at a specific frequency from one height to another. The frequency corresponding to maximum ankle-section current or SAR would, however, increase as $1/h$, to being on the order of 50.7 MHz for a 10-year-old (h = 1.38 m) and 62.5 MHz for a 5-year-old (h = 1.12 m), as against 40 MHz for an adult. Since the current increases as f, the maximum SARs projected at the new peak SAR frequencies would be considerably higher. Linearly interpolating between the values given at 40.68 and 100 MHz (Johnson and Guy, 1972), we have taken larger conductivities σ_c of the wet tissues at the higher frequencies and have used the values 0.73 and 0.77 S/m at 50.7 and 62.5 MHz, respectively. The highest ankle-section SARs projected for 10- and 5-year-old children for 1 mW/cm^2 incident plane waves (E = 61.4 V/m) are estimated to be 371 and 534 W/kg, respectively. The corresponding ankle-section current densities are 52 and 64 mA/cm^2 for ten- and five-year-old children, respectively.

CONTACT HAZARDS IN THE VLF TO HF FREQUENCY BAND

The problem of RF shock and burns for human contact with commonly encountered metallic bodies has been discussed in the literature (Gandhi and Chatterjee, 1982; Guy and Chou, 1982; Rogers, 1981). Ungrounded objects such as car, van, bus, fence, metallic roof, etc., will develop open-circuit voltages on the order of several hundred volts when exposed to the ANSI-recommended electric fields of 614 V/m for the frequency band 0.3 to 3 MHz. We have recently completed a study where the body impedance and threshold currents needed to produce sensations of perception and pain were measured for 367 human subjects (197 male and 170 female) for the frequency range 10 kHz to 3 MHz (Chatterjee et al., 1986). The study included various types of contact such as finger contact and grasping a rod electrode to simulate holding a car's door handle. Based on the principles enunciated in an earlier paper (Gandhi and Chatterjee, 1982), Chatterjee et al. (1986) have used the new data to estimate the incident electric fields that will cause perception upon holding the door handle of various vehicles such as a compact car, van, or school bus. These values are shown in Fig. 3 for an adult male and scaled therefrom for a 10-year-old child. The sensation is one of tingling or pricking for frequencies less than about 100 kHz and warming for higher frequencies. Currents on the order of 250 mA are needed to cause perception for a grasping contact. Although currents necessary to induce pain could not be obtained from the RF generators available for this study, Chatterjee et al. (1986) estimate the same to be about 20 percent higher, based on the experience with finger contact measurements. The threshold electric fields given in Fig. 3 may therefore be multiplied by a factor of 1.2 to obtain an estimate of incident E-fields that will cause pain upon holding the handle.

THERMAL IMPLICATIONS OF HIGH SARs

From the foregoing it is obvious that substantial current densities and concomitant SARs may be set up in the body extremities such as the ankle section for freestanding individuals (no contact with metallic bodies) or the wrist section for conditions of contact with the metallic bodies, for electric-field limits suggested in the ANSI C95.1-1982 RF safety guideline. We have measured the surface temperature elevation of the wrist and the ankle sections for a healthy human subject at room temperature for a variety of RF currents and

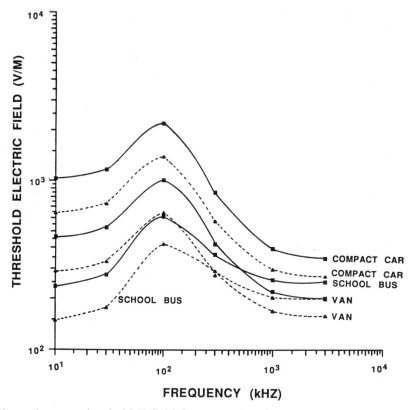

Fig. 3. Average threshold E-field for perception for grounded adult males (solid curves) and ten-year-old children (dashed curves) in grasping contact with various metallic objects.

estimated SARs in the frequency band 1-50 MHz (Chen and Gandhi, 1988). A couple of representative graphs on the temperature of several locations of the ankle and wrist sections are given as Figs. 4 and 5. As a result of the 75+ experiments that have been performed, the observed highest rates of temperature increase $\Delta T/\Delta t$ in °C/min (for point 1 of the ankle section -- see Fig. 4, for legend and for point a of the wrist section -- see Fig. 5) are given by the following best-fit relationships:

$$\frac{\Delta T}{\Delta t}\bigg|_{\text{ankle section}} = 0.0045 \times \text{SAR} \quad °C/\text{min}$$

$$\frac{\Delta T}{\Delta t}\bigg|_{\text{wrist section}} = 0.0048 \times \text{SAR} \quad °C/\text{min}$$

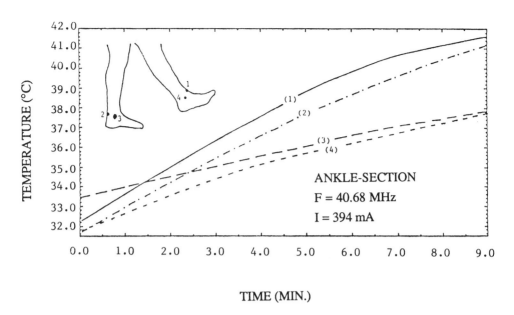

Fig. 4. Temperature of several locations at the surface of the ankle section under condition of RF current through the leg. Estimated SAR in high-water-content tissues = 239 W/kg.

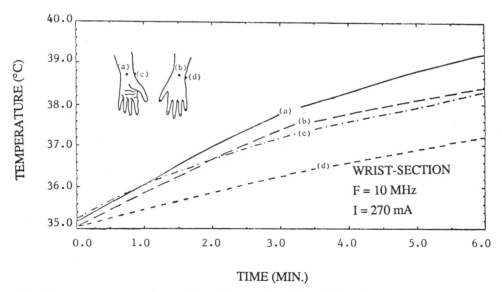

Fig. 5. Temperature of several locations at the surface of the wrist section under condition of RF current through the hand. Estimated SAR in high-water-content tissues = 134 W/kg.

where the SAR is in W/kg. Since ankle-section SARs on the order of 182-243 W/kg and wrist-section SARs as high as 1045 W/kg have been projected previously for the E-fields recommended in the ANSI C95.1-1982 safety guide, fairly high rates of temperature increase are therefore anticipated. This is also obvious from the data given in Figs. 4 and 5.

COUPLING OF THE HUMAN BODY TO RF MAGNETIC FIELDS

As pointed out earlier the vertically polarized RF electric field associated with plane waves is capable of inducing fairly significant RF currents in the human body, particularly for conditions of contact with the ground. Leading as it does to fairly high SARs in the ankle region, this observation may lead to a future restriction of the RF electric fields in the safety standards in the HF region of frequencies. For example, maximum electric-field magnitudes as low as 27.5 V/m have been proposed for the public exposure limit for the frequency band 10-400 MHz (IRPA, 1988). In the past, the safety standards have generally allowed a magnetic field that was 1/377 of the electric field, in MKS units. It is therefore of interest to examine the coupling of the RF magnetic field with the human body to see if a commensurate restriction of the magnetic-field component will be warranted. Toward this objective we have used the 3-D impedance method to calculate the induced currents and the corresponding SARs in the human body where the magnetic field is polarized from front to back (Orcutt and Gandhi, 1988). This orientation was selected because of its strongest coupling to the human body.

We used the inhomogeneous, anatomically-based model of the human body described by Sullivan et al. (1988). Using 1.31 cm cubic cells, this model specifies the conductivity and dielectric constant for each of about 144,000 cells in a 59 x 31 x 175 cm prism. About 64,000 of these cells are entirely in air. Figure 6 shows the calculated layer-averaged SARs for the model of the human body exposed to a magnetic field of 1 Ampere/meter at 30 MHz, polarized from the front to back of the model. The whole-body-averaged SAR is 0.03 W/kg. The peak SAR for this case, 0.64 W/kg, occurred in layer number 72.

SUGGESTED RADIO-FREQUENCY PROTECTION GUIDE
(RFPG) FOR OCCUPATIONAL EXPOSURES

To reduce the abovementioned problems, new limits have recently been proposed by IRPA (1988) and by Dr. Maria Stuchly of Health & Welfare, Canada (1987). Using the recent data, we propose the following modifications of the radio-frequency protection guides to limit the RF currents that can be induced in the human body.

The proposed RFPG is given in Table 1 and plotted in Fig. 7.

Since higher E-fields proposed in Table 1 for the band 0.003-100 MHz, if these were vertical, would result in high RF-induced body currents and a potential for shock and burns for contact with ungrounded metallic bodies, the personnel access areas should be limited in the following manner:

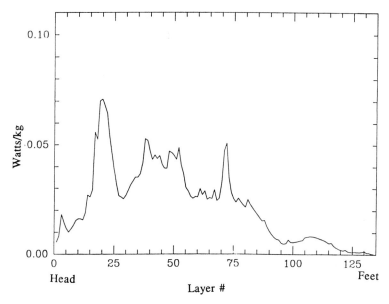

Fig. 6. Layer-averaged SARs for the model of the human body, when the body is exposed to a linearly polarized RF magnetic field of 1 A/m at 30 MHz, polarized from front to back. Each of the layers is 1.31 cm in thickness.

Table 1. Proposed Radio-frequency Protection Guides for Occupational Exposures

Frequency Range (MHz)		E V/m	H A/m	Plane-Wave Equivalent Power Density (mW/cm^2)
0.003 - 0.1	*	614	163	---
0.1 - 3.0	*	614	16.3/f	---
3 - 30	*	1842/f	16.3/f	---
30 - 100	*	61.4	16.3/f	---
100 - 300		61.4	0.163	1.0
300 - 3000		61.40x(f/300)$^{1/2}$	0.163x(f/300)$^{1/2}$	f/300
3000 - 300,000		194	0.5	10.0

Note: f = frequency in MHz. E and H are the magnitudes of electric and magnetic fields, respectively.

* The personnel access areas should be restricted to limit induced RF body currents and potential for RF shock and burns, as defined in the following.

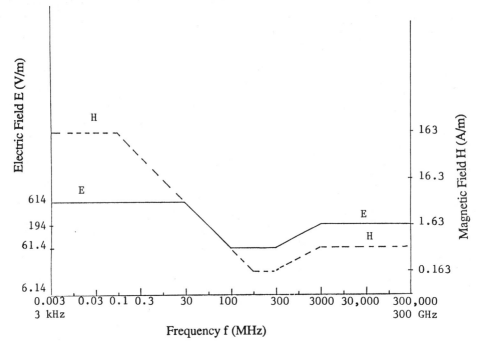

Fig. 7. Proposed radio-frequency protection guide for occupational exposures.

1. For free-standing individuals (no contact with metallic bodies), RF current induced in the human body should be less than or equal to 100 mA as measured through both feet or 50 mA through each of the feet. For a frequency of less than 100 kHz, the allowable induced current should be reduced as follows:

$$I = 1.0\, f_{kHz} \quad mA \tag{6}$$

The above limitations on RF induced currents are suggested to ensure that the ankle-section SARs for frequencies higher than 0.1 MHz will be no more than 5.8-10.7 W/kg for adults of heights 1.75-1.5 m. For frequencies lower than 100 kHz, the current densities in the ankle section will be slightly lower than those needed for stimulating thresholds for the nerve/muscle system (Chatterjee et al., 1986; Berhnardt, 1985; Gandhi et al., 1985b; Guy and Chou, 1985).

For vertically polarized electric fields, the above limitation on current would imply (Gandhi et al., 1986) that the permissible E-field is less than (300/f) V/m for frequencies in excess of 0.1 MHz.

2. For conditions of contact with metallic bodies, maximum RF current through an impedance equivalent to that of the human body for conditions of grasping contact (see Chatterjee et al., 1986 -- Fig. 1), as measured with a contact current meter shall not exceed the following values:

$$I = 0.5 f_{kHz} \quad \text{mA}$$

for $3 \leq f \leq 100$ kHz (7)

$$= 50 \text{ mA}$$

for $f > 0.1$ MHz (8)

The current limits given by Eqs. (7) and (8) would help ensure that the current experienced by a human being upon contacting these metallic bodies would be less than that needed for perception or pain at each of the frequencies.

Steps such as grounding and use of safety equipment that result in reduced currents would obviously allow existence of higher fields without exceeding the above limits for conditions of contact with metallic bodies. Significantly higher RF magnetic fields are recommended in the proposed RFPG of Table 1. For the frequency band 0.1-100 MHz, the RF magnetic-field guideline is

$$H = \frac{16.3}{f} \quad \text{A/m} \tag{9}$$

For magnetic fields given by Eq. (9), the peak and whole-body-averaged SARs have been estimated using the data of the previous section and Fig. 6. These are given in Table 2 along with the peak internal current densities. A magnetic-field orientation from front to back of the body is assumed for these calculations. This orientation was selected because of its strongest coupling to the human body.

For frequencies less than 0.1 MHz, an RF magnetic field of 163 A/m implies a peak current density $\leq 0.0117 f_{kHz}$ mA/cm^2 which is considerably lower than the threshold of perception of currents at these frequencies.

SUGGESTED RFPG FOR THE GENERAL PUBLIC

The proposed RFPG for the general public is given in Table 3. This RFPG is plotted in Fig. 8. The explanations for superscripts a-d are given in the following:

a. The electric field E suggested for these frequency bands is lower than the threshold of perception of commonly encountered metallic bodies such as a car, a van, etc. It is, however, close to the threshold of perception for finger contact of a school bus by a child (Chatterjee et al., 1986).

b. For the E-field suggested here, the current induced in a freestanding (no contact with metallic bodies) human being is less than or equal to 100 mA (one leg current = 50 mA), which is consistent with the access area limitation for occupational exposures.

c. An incident electric field of 8 V/m implies the numbers given in Table 4 for maximum induced currents and ankle-section SARs.

d. We have previously projected that whole-body exposure millimeter-wave power densities on the order of 8.7 mW/cm^2 are likely to cause sensations of "very warm to hot" (Gandhi and Riazi, 1986). At higher frequencies, a power density of 1 mW/cm^2 is suggested to prevent threshold of perception of warmth.

Table 2. Whole-body-averaged and peak SARs for an anatomically-realistic model of an adult human being for RF magnetic fields given by Eq. (9). A magnetic-field orientation from front to back of the body is assumed to obtain highest possible SARs.

f_{MHz}	H A/m	Whole-body-averaged SAR W/kg	Peak Current Density mA/cm^2	Peak SAR W/kg
0.1	163*	0.014	1.17	0.31
0.3	54.3	0.013	1.17	0.29
1.0	16.3	0.012	1.17	0.27
3.0	5.43	0.011	1.17	0.24
10.0	1.63	0.010	1.17	0.22
30.0	0.54	0.009	1.17	0.20
100.0	0.16	0.006	1.17	0.13

* A magnetic field of 163 A/m corresponds to a magnetic flux of 2.05 Gauss in the Gaussian system of units.

Table 3. Proposed Radio-frequency Protection Guides for general population

Frequency Range (MHz)	E** V/m		H** A/m	Plane-Wave Equivalent Power Density (mW/cm^2)	
0.003 - 0.1	(a)	61.4*	163	---	
0.1 - 1.0	(a)	61.4*	16.3/f	---	
1.0 - 3.9	(b)	61.4†	16.3/f	---	
3.9 - 100	(c)	240/f†	16.3/f	---	
30 - 100		8†	16.3/f	---	
100 - 5,900		0.8 f$^{1/2}$†	0.163	---	
5,900 - 300,000		61.4	0.163	(d)	1.0

* spatially-averaged over a volume corresponding to that of an automobile.

† spatially-averaged over a volume corresponding to that of a human being.

** The frequency f is in MHz. E and H are the magnitudes of electric and magnetic fields, respectively. Explanation for (a)-(d) is given in the text.

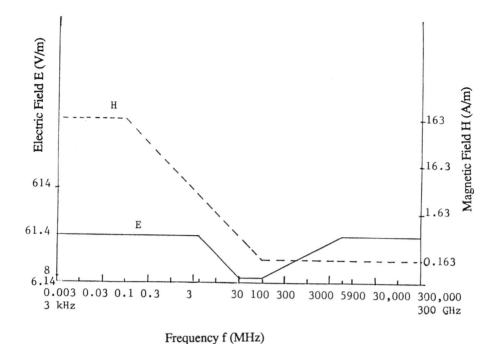

Fig. 8. Proposed radio-frequency protection guide for general population.

Table 4. Induced currents and ankle-section SARs for an incident E-field of 8 V/m (Gandhi et al., 1986)

	Height m	Frequency MHz	I_h mA	Ankle SAR* W/kg
Average adult	1.75	40.0	101.6	3.8
Adult	1.5	46.7	87.2	5.0
10-year-old child	1.38	50.7	80.1	8.7
5-year-old child	1.12	62.5	65.0	7.8

* We have assumed a conductivity of 0.693 S/m for the high-water-content tissues at 40 and 46.7 MHz (Johnson and Guy, 1972). Somewhat larger conductivities of 0.73 and 0.77 S/m are taken at 50.7 and 62.5 MHz, respectively. The corresponding dielectric constants taken are 97.3 for 40 and 46.7 MHz and 92.9 and 87.4 for 50.7 and 62.5 MHz, respectively.

Also the suggested power density of 1 mW/cm^2 is consistent with the recently proposed NCRP guideline for the general population (1986).

Comparison of the Recommended RFPGs with Standards at Other Frequencies

From Eq. (7) the suggested limit on the contact current is 1.5 mA at 3 kHz. This may be compared with the National Electric Safety Code (1977) which specifies a maximum leakage current of 0.5 mA from portable electrical tools and household appliances and 0.75 mA for permanently-fixed appliances. Recognizing the threshold current for perception at 3 kHz is approximately 3 times higher than that at 60 Hz (Dalziel and Mansfield, 1950), a suggested contact current of 1.5 mA is not out of line with the leakage current of 0.5-0.75 mA specified in the National Electric Safety Code.

For higher RF frequencies, the suggested guideline of 10 mW/cm^2 for occupational exposures is in agreement with the occupational standard for infrared radiation, while a reduced guideline of 1 mW/cm^2 (see Table 3) is consistent with the recently proposed NCRP guideline (1986) for the general population.

CONCLUDING REMARKS

The proposed radio-frequency protection guides depart from the previous guidelines such as those by ANSI (1982) in two respects:

1. An increase in the safety levels of RF magnetic fields for frequencies less than 100 MHz. This is proposed in recognition of the fact that these fields do not couple as tightly as do the E-fields and do not cause substantial SARs.

2. Limitation on the induced currents in the human body to no more than 100 mA for frequencies in excess of 100 kHz and a linearly reducing current for lower frequencies. The current under contact situations is limited to 50 mA for frequencies in excess of 100 kHz and a proportionally reducing current at lower frequencies.

Identical current limitations are proposed for both occupational and public exposures. Since safety measures can be adapted in the work place, higher field limits are suggested, provided steps are taken to limit the currents for contact as well as for noncontact situations. Procedures are indeed available to measure the foot currents through the human body using shoe-mounted RF current sensors (Gandhi, 1987). Methods similar to those of Fig. 1 of Chatterjee et al. (1986) can be used to estimate RF currents through the human body for conditions of contact with metallic bodies. For these measurements an impedance equivalent to that of the human body for conditions of grasping contact (Gandhi et al., 1985b) may be used.

REFERENCES

ANSI C2, 1977, "National Electrical Safety Code," Institute of Electrical and Electronics Engineers, Inc., New York.
ANSI C95.1, 1982, "American National Standard -- Safety Levels with Respect to Human Exposure to Radio-frequency Electromagnetic Fields, 300 kHz to 100 GHz," Institute of Electrical and Electronics Engineers, Inc., New York.
Bernhardt, J. H., 1985, Evaluation of human exposures to low frequency fields, in "AGARD Lecture Series No. 138. The Impact of Proposed Radio-frequency Radiation Standards on Military Operations," NATO Advisory Group for Aerospace Research and Development, 7 Rue Ancelle 92200 Neuilly Sur Seine, France.
Chatterjee, I. Wu, D., and Gandhi, O. P., 1986, Human body impedance and threshold currents for perception and pain for contact hazard analysis in the VLF-MF band, IEEE Trans. Biomed. Eng., 33:486.

Chen, J. Y. and Gandhi, O. P., 1988, Thermal implications of high SARs in the body extremities at the ANSI recommended VLF-VHF safety levels, IEEE Trans. Biomed. Eng., 35-435.

Dalziel, C. F., and Mansfield, T. H., 1950, Effect of frequency on perception current, AIEE Trans., :1162.

Durney, C. H., 1980, Electromagnetic dosimetry for models of humans and animals: a review of theoretical and numerical techniques, Proc. IEEE, 68:33.

Durney, C. H. et al., 1978, "Radio-frequency Radiation Dosimetry Handbook," 2nd edition, Report SAM-TR-78-22, USAF School of Aerospace Medicine, Brooks Air Force Base, Texas.

Gandhi, O. P., 1975, Conditions of strongest electromagnetic power deposition in man and animals, IEEE Trans. Microwave Theory and Techniques, 23:1021.

Gandhi, O. P., 1979, Dosimetry -- The absorption properties of man and experimental animals, Bull. NY Acad. Medicine, 55:999.

Gandhi, O. P., 1987, "RF Personnel Dosimeter and Dosimetry Method for Use Therewith," US Patent No. 4,672,309.

Gandhi, O. P. and Chatterjee I., 1982, Radio-frequency hazards in the VLF to MF band, Proc. IEEE, 70:1462.

Gandhi, O. P., Chatterjee, I., Wu, D., and Gu, Y. G., 1985a, Likelihood of high rates of energy deposition in the human legs at the ANSI recommended 3-30 MHz RF safety levels, Proc. IEEE, 73:1145.

Gandhi, O. P., Chatterjee, I., Wu, D., D'Andrea, J. A., and Sakamoto, K., 1985b, "Very Low Frequency Hazard Study," report prepared for USAF School of Aerospace Medicine, Brooks Air Force Base, Texas, on Contract F-33615-83-R-0613.

Gandhi, O. P., Chen, J. Y., and Riazi, A., 1986, Currents induced in a human being for plane-wave exposure conditions 0-50 MHz and for RF sealers, IEEE Trans. Biomed. Eng., 33:757.

Gandhi, O. P., Hunt, E. L., and D'Andrea, J. A., 1977, Deposition of electromagnetic energy in animals and in models of man with and without grounding and reflector effects, Radio Science, 12(6S):39.

Gandhi, O. P. and Riazi, A., 1986, Absorption of millimeter waves by human beings and its biological implications, IEEE Trans. Microwave Theory and Tech., 34:228.

Gronhaug, K. L. and Busmundrud, K. L., "Antenna Effect on the Human Body of EMP," (in Norwegian), Report FFI/NOTAT-82/3013, Norwegian Defense Research Establishment, Kjeller, Norway.

Guy, A. W., 1987, Measured body to ground current and thermal consequences for human subjects exposed to 3.68 MHz to 144.5 MHz radio-frequency fields, Abstracts of the Ninth Ann. Meet. of BEMS, Portland, Oregon, :31.

Guy, A. W. and Chou, C. K., 1982, Hazard analysis: Very low frequency through medium frequency range, Final Report on USAF SAM Contract F-33615-78-D-0617 Task 0065.

Guy, A. W. and Chou, C. K., 1985, "Very Low Frequency Hazard Study," final report prepared for USAF School of Aerospace Medicine, Brooks Air Force Base, Texas, on Contract F-33615-83-C-0625.

Hill, D. A., 1984, The effect of frequency and grounding on whole-body absorption of humans in E-polarized radio-frequency fields, Bioelectromagnetics, 5:131.

Hill, D. A. and Walsh, J. A., 1985, Radio-frequency current through the feet of a grounded human, IEEE Trans. Electromag. Comp., 27:18.

IRPA, 1984, Interim guidelines on limits of exposure to radiofrequency electromagnetic fields in the frequency range from 100 kHz to 300 GHz, Health Physics, 46:975.

IRPA, 1988, Guidelines on limits of exposure to radio-frequency electromagnetic fields in the frequency range from 100 kHz to 300 GHz, Health Physics, 54:115.

Johnson, C. C. and Guy, A. W., 1972, Nonionizing electromagnetic wave effects on biological materials and systems, Proc. IEEE, 60:692.

Morton, D. J. Truex, R. C., and Kellner, C. E., 1941, "Manual of Human Cross Section Anatomy," The Williams and Wilkins Company, Baltimore.

National Council on Radiation Protection and Measurements, 1986, "Biological Effects and Exposure Criteria for Radiofrequency Electromagnetic Fields," NCRP Report No. 86, Bethesda, Maryland.

Orcutt, N., and Gandhi, O. P., 1988, A 3-D impedance method to calculate power deposition in biological bodies subjected to time-varying magnetic fields, IEEE Trans. Biomed. Eng., 35:577.

Rogers, R. J., 1981, Radio-frequency burn hazards in the MF/HF band, in "Aeromedical Review -- Proceedings Workshop on the Protection of Personnel Against RF Electromagnetic Radiation" J. C. Mitchell, ed., Brooks Air Force Base, Texas.

Stuchly, M. A., 1987, Proposed revision of the Canadian recommendations on radio-frequency exposure protection, Health Physics, 53:649.

Sullivan, D., Gandhi, O. P., and Taflove, A., 1988, Use of the finite-difference time-domain method in calculating EM absorption in man models, IEEE Trans. Biomed. Eng., 35:179.

Tell, R., Mantiply, E. D., Durney, C. H., and Massoudi, H., Electric and magnetic field intensities and associated induced body currents in close proximity to a 50 kW AM standard broadcast station, to be published in IEEE Trans. Broad..

INTERACTION MECHANISMS AT MICROSCOPIC LEVEL

Paolo Bernardi and Guglielmo D'Inzeo

Department of Electronics
University of Rome "La Sapienza"
Via Eudossiana 18, 00184 Rome, Italy

INTRODUCTION

The interaction mechanisms can be examined at different organization levels of the biological matter. The study can be developed considering complex systems, as organs and tissues, then less complex ones, as chains or groups of cells, and ultimately cells or subcellular components (membrane, nucleus, membrane channels). The lowest subcellular level gives the deepest knowledge on the interaction characteristics and allows an analysis of the physical and chemical action induced by the EM field on the charges of the biological body.

At subcellular level the interaction essentially develops through the forces that the electromagnetic field exerts on the free and bounded electric charges of the biological system. As a consequence, the interaction's characteristics are linked to the type of charge and its properties. It is important to outline that while the forces acting on the charges depend only on the EM field and on the electrical characteristics of the particle, the dynamics of these particles (vibrations, displacements, rotations) also depends on other properties (i.e. the mass, the kind of particle: atomic or molecular, the chemical bonds, etc.).

The interaction mechanisms at microscopic level essentially depend on the type of the involved charged particles, say unipolar charges (positive or negative) or dipolar charges (total neutral charge). In biological systems the first ones are mainly ions, while the latter are essentially water molecules and neutral proteins. Atoms and molecules with both kinds of charges can be present in a biological system; for example, acid and basic proteins can have both dipolar and unipolar characteristics.

In the following we divide the actions on the charges in:
(a) displacements of free charges (mainly ions) from the resting position;
(b) increases of vibrations in bounded charges (electrons in atoms, atoms in molecules);

(c) rotation and reorientation of dipolar molecules (mainly molecules of water and neutral proteins).

MACROSCOPIC INTERACTION: THE DIELECTRIC POLARIZATION AND THE RELAXATION TIME

The actions produced by the EM field on unipolar particles and dipolar molecules are respectively:
a force $\boldsymbol{F}$ on an electric charge q

$$\boldsymbol{F} = q(\boldsymbol{E} + \boldsymbol{v} \times \boldsymbol{B}) \qquad (1)$$

where $\boldsymbol{v}$ is the charge's velocity;
a torque $\boldsymbol{T_e}$ on an electric dipole

$$\boldsymbol{T_e} = \boldsymbol{p_e} \times \boldsymbol{E} \qquad (2)$$

where $\boldsymbol{p_e} = q\boldsymbol{l_0}$ is the dipole moment. This torque tends to align the dipole in the direction of the applied field.

Considering the biological system at macroscopic level, the effects of the previous actions on atoms and molecules can be represented by the dielectric polarization vector $\boldsymbol{P}$ defined as the induced electric dipole moment per unit volume of the considered material. The electric induction $\boldsymbol{D}$ in the body is therefore:

$$D = \varepsilon_0 \boldsymbol{E} + \boldsymbol{P} = \varepsilon_0 \varepsilon_r \boldsymbol{E} \qquad (3)$$

and

$$\boldsymbol{P} = \varepsilon_0 (\varepsilon_r - 1)\boldsymbol{E} = \varepsilon_0 \chi \boldsymbol{E} \qquad (4)$$

where ε_0 is the permittivity of vacuum, ε_r the relative permittivity, χ the electric susceptibility of the material.

The rate of the induced polarization is limited by the dipolar particle reorientation and by the atomic polarization. The slowest polarization mechanism is the dipolar reorientation, so that it is the first polarization term that disappears as the frequency of the applied field increases. The atomic polarization is much faster so that the fall-off of this contribution occurs at frequencies of the order of the natural frequencies of vibration of the atoms in a molecule. We can therefore study the two polarization effects separately, say the dipolar reorientation and the atomic polarization, with two different models.

The model for the dipolar reorientation assumes that the dielectric polarization $\boldsymbol{P}$ is due to the sum of two contributions: $\boldsymbol{P_1}$ and $\boldsymbol{P_2}$. The polarization $\boldsymbol{P_1}$ arises from atomic and electronic displacements and responds instantly to the applied electric field $\boldsymbol{E}$:

$$\boldsymbol{P_1} = \varepsilon_0 \chi_1 \boldsymbol{E} \qquad (5)$$

while $\boldsymbol{P_2}$ arises from dipolar reorientation and delays with respect to $\boldsymbol{E}$ approaching the instantaneous response $\varepsilon_0 \chi_2 \boldsymbol{E}$ at a rate proportional to $\varepsilon_0 \chi_2 E - P_2$, so that:

$$\frac{dP_2}{dt} = \frac{1}{\tau}(\varepsilon_0 \chi_2 E - P_2) \tag{6}$$

The dielectric polarization P deriving from an E-field step of amplitude E_M (i.e. $E(t) = E_M u_{-1}(t)$), applied at $t = 0$, when $P_2 = 0$, is

$$P = P_1 + P_2 = \varepsilon_0 [\chi_1 + \chi_2 (1 - \exp(-t/\tau))] E_M \tag{7}$$

which shows that the final value is reached exponentially with a time constant τ called relaxation time. If the applied field is time-harmonic with an angular frequency ω, the dielectric polarization is

$$P = \varepsilon_0 \left(\chi_1 + \frac{\chi_2}{1+j\omega\tau} \right) E \tag{8}$$

and therefore:

$$D = \varepsilon_0 E + P = \varepsilon_0 \left(1 + \chi_1 + \frac{\chi_2}{1+j\omega\tau} \right) E = \varepsilon_0 (\varepsilon' - j\varepsilon'') E \tag{9}$$

Equation (9) defines the complex relative permittivity $\varepsilon' - j\varepsilon''$ whose dispersion formula is

$$\varepsilon' - j\varepsilon'' = \varepsilon_\infty + \frac{\varepsilon_s - \varepsilon_\infty}{1+j\omega\tau} \tag{10}$$

where ε_∞ is the permittivity at sufficiently high frequency as to the disappearing of dipolar reorientation, while ε_s is the permittivity at low frequency when all the dipolar molecules can follow the variations of the applied field. Equation (10) is commonly known as the Debye relaxation formula.

The second polarization effect, the atomic polarization, can be studied assuming the model of an atom with a nucleus of charge $+q$ surrounded by a spherically symmetrical electron cloud $-q$. The application of an E-field displaces these charges inducing a dipole moment (Fig.1). The polarization P produced by this mechanism must satisfy the equation [Collin, 1966]

$$\frac{d^2 P}{dt^2} + v \frac{dP}{dt} + \omega_p^2 P = \frac{Nq^2}{m} E(t) \tag{11}$$

where m is the mass of the electron cloud, N is the number of polarized atoms per unit volume, and v and ω_p are characteristic of the particular biological medium.

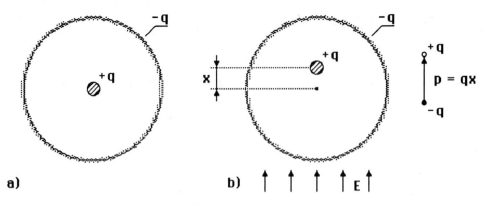

Fig.1. Model of an atom with a nucleus +q surrounded by a spherically symmetrical electron cloud -q.

In the time domain the response to an electric field step of amplitude E, applied at $t=0$, is

$$P(t) = \frac{Nq^2 E}{m\omega_p^2}\left(1 + \frac{p_2}{p_1 - p_2}e^{p_1 t} - \frac{p_1}{p_1 - p_2}e^{p_2 t}\right) \quad (12)$$

where p_1 and p_2 are the roots of the characteristic equation of Eq.(11).

In the frequency domain the dielectric polarization is:

$$P = \frac{Nq^2}{m(\omega_p^2 - \omega^2 + j\omega\nu)} E \quad (13)$$

and the complex relative permittivity is given by:

$$\varepsilon' - j\varepsilon'' = 1 + \frac{Nq^2}{\varepsilon_0 m(\omega_p^2 - \omega^2 + j\omega\nu)} \quad (14)$$

From (13) it follows that the charges have a proper frequency of oscillation, ω_0, given by:

$$\omega_0 = \sqrt{\omega_p^2 - \left(\frac{\nu}{2}\right)^2} \quad (15)$$

and the power loss will be expected to have a maximum near this frequency.

MICROSCOPIC INTERACTION: THE MOLECULAR POLARIZABILITY AND THE LOCAL FIELD

Molecules can be divided into two classes, polar and non polar, according to whether or not they possess an intrinsic electric dipole moment in their lowest energy-level. The magnitude of molecular dipole moments is usually of the order of the electronic charge ($q = 1.6 \times 10^{-19}$ C) displaced by interatomic distances, say $0.25 \cdot 10^{-10}$ m, that is $4 \cdot 10^{-30}$ Cm or about 1 D (1 Debye = $3.33 \cdot 10^{-30}$ Cm). The purpose of this section is to determine the properties of dielectric substances consisting of molecules having an intrinsic dipole moment μ, when free of perturbing influences.

In addition to its translational motion a free molecule can carry out oscillations and rotations. However, it is assumed that these do not alter the average value of the dipole moment. An electric field E_1 influences the molecule in two ways. First, it perturbs the free rotation of the dipole and, second, it introduces a further dipole moment, say αE_1, by elastic displacement of the atomic electrons relative to their nuclei (Fig.2) and by elastic displacement of the nuclei relative to each other (Fig.3). The total moment of the molecule is thus

$$m = \mu + \alpha E_1 \qquad (16)$$

The quantity α is called the polarizability of the molecule and is given by the sum of two terms

$$\alpha = \alpha_e + \alpha_a \qquad (17)$$

where α_e is the electronic polarizability   (Fig. 2),
        α_a is the atomic polarizability   (Fig. 3).

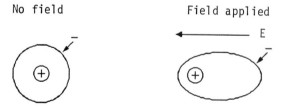

Fig.2. Schematic representation of the electronic polarization.

Fig.3. Schematic representation of the atomic polarization.

In Eq.(4) the polarization vector **P** has been defined in a macroscopic form. Alternatively, the polarization **P** is equivalent to the electric dipole moment per unit volume of the material. This latter interpretation provides a link between macroscopic and molecular points of view. In fact, the dipole moment per unit volume results from the additive action of N elementary average (overline) dipole moments:

$$\mathbf{P} = N \overline{\mathbf{m}} = N(\overline{\mu} + \alpha \mathbf{E}_1) \tag{18}$$

where N is the number of dipoles per unit volume, and $\overline{m}$ is the average value of the dipole moment under the action of the local field $\mathbf{E}_1$. When an electric field is applied to a material composed of dipolar molecules, the dipoles experience a torque (Eq. 2) tending to align them with the field so that:

$$\overline{\mu} = \alpha_d \mathbf{E}_1 \tag{19}$$

where α_d is called orientation (or dipolar) polarizability (Fig.4).

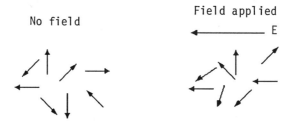

Fig.4. Schematic representation of the dipolar polarization.

From Eqs.(17-19) it follows that:

$$\mathbf{P} = N (\alpha_d + \alpha_e + \alpha_a) \mathbf{E}_1 \tag{20}$$

and therefore, from (4) and (20):

$$\varepsilon_0 (\varepsilon_r - 1) \mathbf{E} = N (\alpha_d + \alpha_e + \alpha_a) \mathbf{E}_1 = N \alpha_T \mathbf{E}_1 \tag{21}$$

where α_T is called the total polarizability of the molecule.
It should be noticed that, in general, in Eq.(21) (that relates macroscopic to molecular quantities) the microscopic local field $\mathbf{E}_1$ acting on a molecule has to be distinguished from the macroscopic field **E** existing in the biological sample. It is the goal of the molecular theories to evaluate these parameters and to quantitatively relate the macroscopic permittivity ε_r and the total polarizability α_T for a given material.

For dilute solutions of dipolar molecules in a non-polar solvent and for pure liquids of low polarity the Clausius-Mossotti-Lorentz theory (von Hippel, 1954) gives:

$$\mathbf{E}_1 = \frac{\varepsilon_r + 2}{3} \mathbf{E} \tag{22}$$

and therefore, from (21):

$$\frac{\varepsilon_r - 1}{\varepsilon_r + 2} = \frac{N\alpha_T}{3\varepsilon_0} \qquad (23)$$

An interesting consequence of (22) is that, if a biologic spherical specimen with these characteristics is brought into a uniform external field $\mathbf{E_e}$ in air, this field also represents the local field on each molecule. In fact, the macroscopic field inside a spherical specimen of relative permittivity ε_r exposed to $\mathbf{E_e}$ is uniform and equal to:

$$\mathbf{E} = \frac{3}{\varepsilon_r + 2} \mathbf{E_e} \qquad (24)$$

and hence, taking (22) into account, $\mathbf{E_1} = \mathbf{E_e}$.

From (23) it follows that for a biological material having a dispersion due to dipolar relaxation the low frequency permittivity, ε_s, is given by

$$\frac{\varepsilon_s - 1}{\varepsilon_s + 2} = \frac{N}{3\varepsilon_0}\left(\alpha_d + \alpha_e + \alpha_a\right) \qquad (25)$$

while, at frequencies so high that the dipoles cannot attain equilibrium, the permittivity ε_∞ is given by

$$\frac{\varepsilon_\infty - 1}{\varepsilon_\infty + 2} = \frac{N}{3\varepsilon_0}\left(\alpha_e + \alpha_a\right) \qquad (26)$$

Rearranging (25) and (26) we obtain:

$$\frac{\varepsilon_s - \varepsilon_\infty}{(\varepsilon_s + 2)(\varepsilon_\infty + 2)} = \frac{N\alpha_d}{9\varepsilon_0} \qquad (27)$$

This equation can be used to obtain quite accurate values for the dipolar polarizability from measurement of the dielectric dispersion.

The previous results are based on the assumption that the dipolar interaction energy is small compared with the thermal energy ($\cong kT$ per molecule). This is not true for liquids in which the hydrogen atoms form bonds among the molecules producing short-range forces as in water and in hydroxylic compounds.

DIPOLAR STRUCTURE OF WATER AND PROTEIN MOLECULES

In order to understand the dielectric properties of water in biological materials it is convenient to consider first the dielectric behavior of pure water. The pure water macroscopic behavior is characterized by a single Debye dispersion term with a permittivity decrease $\varepsilon_s - \varepsilon_\infty$ of about 75 at a relaxation frequency f_τ which depends on the temperature (e.g. about 18.7 GHz at 293 K, Fig. 5).

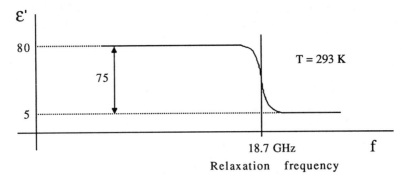

Fig.5. Dispersion curve of pure water.

The molecular interpretation of the relaxation time τ was given first by Debye for polar molecules. The dipolar molecules rotate under the action of the torque T of the applied field with an angular velocity proportional to the torque:

$$T = \xi \frac{d\theta}{dt} \qquad (28)$$

The molecular friction constant ξ depends on the shape of the molecule and on the interaction it encounters. Debye considers dipolar molecules as spheres whose rotation is opposed by the viscosity of the surrounding medium. With this assumption, Stokes' classical hydrodynamic calculations give:

$$\xi = 8\pi\eta a^3 \qquad (29)$$

where a is the dipolar sphere radius and η the macroscopic viscosity of the fluid. Debye was able to calculate the relaxation time τ statistically by deriving the dipole's space orientation under the counteracting influences of the Brownian motion and of a time-dependent electric field:

$$\tau = \frac{\xi}{2kT} \qquad (30)$$

Combining (29) and (30) it follows that the relaxation time is

$$\tau = \frac{4\pi\eta a^3}{kT} \qquad (31)$$

Taking the molecular radius of water to be one-half the interoxigen distance, $a = 2.8 \cdot 10^{-10}$ m, and the 293 K viscosity $\eta = 10^{-3}$ kg/(m.s), the resulting value of the relaxation time is $\tau = 8.5 \cdot 10^{-12}$ s, in good agreement with the experimental value $\tau = 9.3 \cdot 10^{-12}$ s. The corresponding relaxation frequency is $f_\tau = 18.7$ GHz. The fact that the rotating sphere model gives good results for the relaxation time is rather surprising since the water dipolar molecules have a behavior far from that of spheres in a non-polar fluid, in view of the strong hydrogen bonds among them.

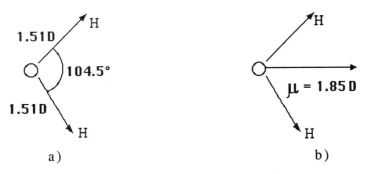

Fig.6. Isolated water molecule's dipole moment.

Intermolecular hydrogen bonding plays a prominent role in maintaining the water's stable structure and its unique electrical properties. These strong intermolecular forces arise from the particular distribution of electrons in the water molecule. Each of the hydrogen atoms shares a pair of electrons with the oxygen atom making an OH dipole moment of 1.51 D (Fig.6a). The HOH bond angle is 104.5°, so that the resultant dipole moment of the water molecule is 1.85 D (Fig. 6b). The formation of hydrogen bonds in water is accompanied by a redistribution of electronic charges, with the H-atom in the bond losing electronic charge and the oxygen atoms gaining it (Fig 7b). Loss of positive charge occurs for the hydrogen atoms attached to the proton-donor molecule, while the hydrogen atoms attached to the proton-acceptor molecule gains positive charge upon the formation of the hydrogen bond (Fig 7c).

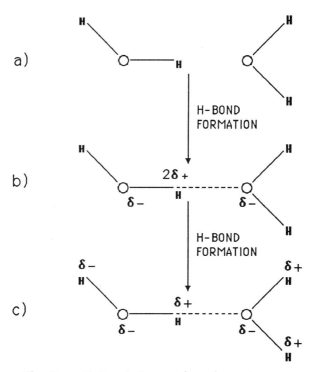

Fig.7. H-Bond formation in water.

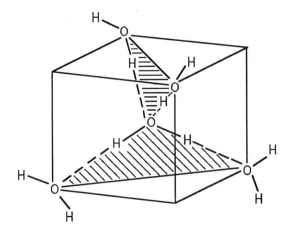

Fig.8. Tetrahedral coordination of water molecules.

An improved model for the dipolar polarization in water is obtained considering the dipole molecule with its first layer of neighbors as a molecular island floating in a continuum in which hydrogen bonds are broken and reformed statistically. With this model good results have been obtained by Kirkwood (Fröhlich, 1950) considering the first shell of neighbors in tetrahedral coordination, as shown in Fig.8.

Dipole moments of protein molecules range in the order of 100 to 1000 D. These large moments derive from the great length of these molecules and not by charges higher than in normal molecules. For example, the peptide unit of a protein (Fig. 9) has a permanent dipole moment of about 3.6 ÷ 3.7 D (Pethig, 1979).If the peptide backbone is in an α-helical configuration with n dipoles, these individual dipoles form a macrodipole (Fig.10a). The resultant dipole is directed along the helix axis with an amplitude equal to the sum of the n dipolar components along this axis. Other forms of peptide structures, e.g. beta sheet, omega turn, random coil have small net dipole strengths.One of the most important biopolymers of life, DNA, is composed by two α-helices.

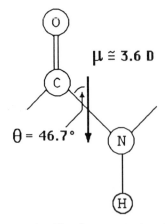

Fig.9. The permanent dipole moment of a peptide unit.

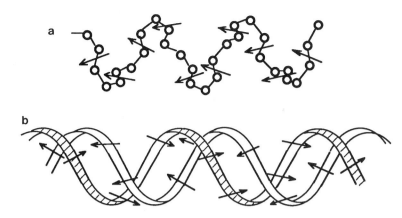

Fig.10. a) Diagram illustrating the way in which the peptide moments add up in an α-helical configuration; b) The double helix of the DNA molecule shows how the peptide moments can cancel each other giving no permanent dipole moment.

Since these two helices point in opposite directions the dipole moments counterbalance exactly, as shown in Fig. 10b. Consequently the DNA macromolecule immersed in an electric field should have a quadrupole behavior.

The model of a sphere rotating in a continuum according to Stoke's law, with a friction factor given by the macroscopic viscosity of the solvent, is particularly appropriate for these giant protein molecules. It is also appropriate for various protein macromolecules in blood serum, such as albumin, hemoglobin, globulin and α and β lipoproteins.

INTERACTION WITH DRIED PROTEINS AND BIOPOLYMERS

The more molecular structures are included in the surroundings of a dipole molecule, the more the dipole itself is anchored in its environment. Hence an activation energy becomes necessary to allow it to orientate freely. In general, a polar molecule in solid dielectrics can assume only determinate positions. The molecular theory of solid dielectrics is based on the two-site model. According to this model the dipoles have only two possible orientations: parallel and antiparallel to the applied electric field. The two orientations are separated by a potential energy barrier per mole U (Fig. 11a). If no electric field is applied, the two sites are taken to be of identical energy and the probabilities P_{12} and P_{21} of transitions per second between the two state are the same. The transition probabilities for no field often are assumed of the form (Pethig, 1979)

$$(P_{12})_0 = (P_{21})_0 = f_0 \exp(-U/kT) \qquad (32)$$

where f_0 is the oscillation frequency in the potential minimum.

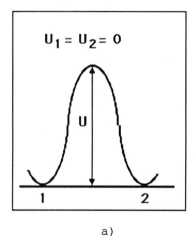 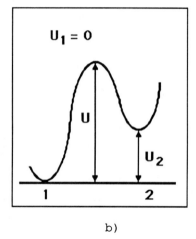

Fig.11. Potential energy barrier: a) Symmetrical double well; b) Asymmetric double well.

On computing the polarization for a sinusoidal electric field, a Debye-type relaxation formula is obtained with relaxation time given by:

$$\tau = \frac{1}{2(P_{12})_0} \qquad (33)$$

and relaxation frequency

$$f_\tau = \frac{f_0}{\pi} \exp(-U/kT) \quad . \qquad (34)$$

In many solids the equilibrium energies of the dipoles are unequal (asymmetric double well, Fig. 11b), and the temperature-dependence of the relaxation frequency is

$$f_\tau = \frac{f_0}{2\pi} (\exp(-U/kT) - \exp(-(U-U_2)/kT)) \qquad (35)$$

The study of the dielectric dispersion of biological macromolecules has been recently extended up to the millimeter and sub-mm wavelengths range (Genzel et al., 1983; Kremer et al., 1984). On the basis of the results obtained, an effort to identify the interaction mechanisms in these frequency ranges has been done. The analysis has been performed with different spectroscopic techniques (Genzel et al., 1983): cavity-perturbation (10 GHz) and untuned-cavity (50 - 150 GHz). The range of temperatures examined was from 4 to 300 K. In these temperature and frequency ranges the absorption coefficient $\alpha = \omega\varepsilon''/nc$ has been measured (n is the real part of the complex index of refraction and c the light velocity in free space). As an example, the results obtained for dried haemoglobin are shown in Fig. 12. Focusing the attention on these results it can be concluded that the observed frequency and temperature dependence cannot be explained with a single Debye relaxation term whose relaxation frequency is assumed to vary according to the theory of the symmetrical double well. The experimental

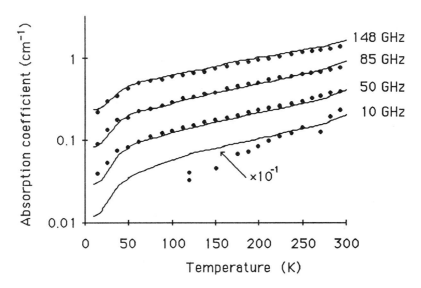

Fig.12. Absorption coefficient of dried haemoglobin from 4 to 300 K. Experimental data (dots) and theoretical model (line).

results (dots, Fig.12) can be fitted with three Debye-type relaxation terms in asymmetrical double well conditions. The parameters for fitting the experimental data of Fig. 12 are given in Table 1 (Kremer et al., 1984).

Table 1. Interaction mechanisms in dried haemoglobin

Relaxation process	High temperature relaxation-frequency limit f_0/π (GHz)	Potential energy barrier U (kcal/mol)	μ (*) (Debye)
1	370	0.29	0.21
2	350	1.11	0.59
3	300	3.60	3.70

(*) μ is the modulus of the difference between the dipole moments in the two positions.

The values in Table 1 lead us to some useful considerations about the microscopic interaction mechanisms. Considering in particular the relaxation process 3, the depth of the potential energy barrier is only slightly below the binding energies of the hydrogen bonds in the NH....OC bridges existing in haemoglobin. Moreover, the NH....OC bridge has a dipole moment of 3.7 D that is the same value of the variation of the dipole moments in the relaxation process. This suggests (Genzel et al., 1983) that process 3 can be a disruption of the hydrogen bond which is changed into a weaker van der Waals bond.

THE CELL MEMBRANE

In the study of the interaction of electromagnetic fields with living systems the main conclusion that can be drawn is that the cell is the primary site of interaction. Most effects cited in the literature refer to alterations of the cell's behavior in terms of changes in its structure, in its response to electrical or chemical stimuli, in the cellular capacity of growing (Adey and Lawrence, 1984; Chiabrera et al., 1985a; Polk and Postow, 1986; Michaelson and Lin, 1987; Lin, 1989). The effects at biological levels more complex than the cellular one can almost always be related to structural modifications or alterations of the communication processes among the cells that form a biological system. On the other hand, some molecular alterations are transferred to cellular level by the cell's capacity of integration and interpretation of biochemical signals.

Some care must be used in the study of the cell behavior. The cell is a complex biological system composed of different parts that can communicate with the outside and also with each other. Figure 13 shows the simplified structure of a cell with the identification of its main components. The understanding of the cell structure is fundamental in the identification of the interaction mechanisms, since several effects are based on a succession of biophysical phenomena that modify the internal activities of the cell.

In the communication between the outside and the inside of the cell the membrane plays a dominant role by controlling its activity. The membrane, while acting as a barrier separating two very different regions, must allow in the meantime the flow of materials through it. Most transport mechanisms through the membrane act by means of its selective permeability to different ions; in these processes various electrical and electrochemical phenomena take place, modifying the transport itself and regulating the cell activity. As an example, let us consider the anesthesia chemical process. In this case the administered drug changes the permeability of the membrane for certain ions, causing an unbalance in the material transport and, as a consequence, an anesthetic effect is induced in some nerves (Milazzo, 1985).In this way, changes in membrane behavior can have macroscopic consequences.

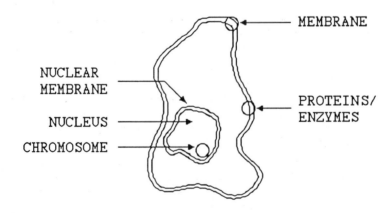

Fig.13. Main components of the cell.

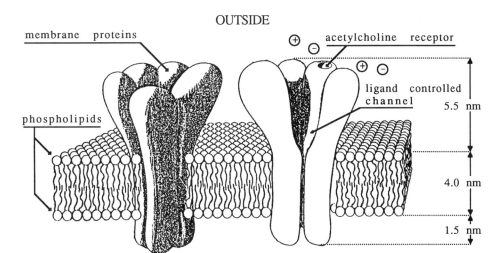

Fig.14. Structural components of the membrane: phospholipids, transmembrane proteins, voltage and ligand controlled channels, surface ions.

In Fig. 14 the main components of the membrane are shown. The phospholipidic bilayer is the barrier that separates the cytoplasm from the outside of the cell; the membrane proteins forming different kinds of membrane channels allow the flow of well determined ions; several ions and chemical substances developing different enzymatic reactions near the inside or the outside of the membrane surface are present. Since most of the cited components, being composed of aminoacid sequences or being solvated ions, are charged or show dipolar properties, the EM field can act on them inducing modifications on their structure and/or on their functionality.

INTERACTION WITH PHOSPHOLIPIDS

In general, a phospholipid is divided in two parts: one head and two tails with similar chemical and physical composition (Fig. 14). It has been recently shown (Seelig et al., 1987) that in some phospholipids the polar heads align themselves nearly parallel (within 30°) to the membrane surface (Fig. 15) and that the heads' group orientation is identical in artificial bilayers and in biological membranes. In Fig. 16a two possible conformations of a phospolipidic bilayer are shown. The transition between one state to the other is affected by the concentration of water at lipid-water interface, and therefore it can be induced isothermally by ionic or solute interactions (Cevc and Marsh, 1987), while, at fixed water concentration, it can be induced increasing the temperature over a critical phase-transition value, T_t (Fig. 16b).

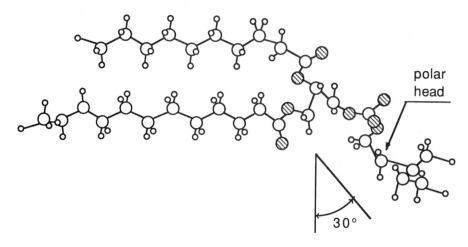

Fig.15. Structure of a phospholipid.

The presence of charges or dipolar molecules externally to the membrane but in the proximity of the heads can modify the bistable nature of the phospholipidic bilayer (Seelig et al., 1987; Cevc and Marsh, 1987). In particular, it has been observed that the phosphocholine head group is sensitive to electric surface charges and that charges of different sign give opposite conformational changes in the membrane (Seelig et al., 1987). Particularly interesting is that relaxation time measurements have shown a rate of reorientation of lipid head-groups at membrane-water interface 1-2 order of magnitude lower than that expected for similar size molecules soluted in water (Shepherd and Büldt, 1978; Scherer and Seelig, 1987); this behavior can depend on the mutual coupling of phospholipidic heads (Seelig et al., 1987) or on the bonds of the heads with the ions present at the membrane interface.

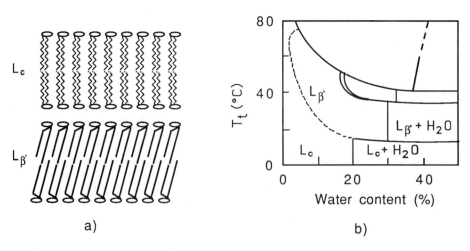

Fig.16. Phase transition of a phospholipidic bilayer: a) Physical structure of two different states; b) phase diagram for a phospholipid.

These recent experimental results can give support to the Bond and Wyeth (1986, 1987) hypothesis that justifies the results of the Liburdy and Penn experiments (1984, 1986). In these experiments the passive sodium ion permeability in red blood cells and in lymphocytes increases under microwave exposure (2.45 GHz, CW, 60 mW/g) in correspondence of the T_t temperature, suggesting that a phase transition can be involved in this microwave effect (Fig. 17). Bond and Wyeth base their theory on the qualitative similarity between the Liburdy and Penn experiments and the change in sodium permeability obtained in vesicles of phospholipids (DPPC) at T_t without microwaves (Papahadjopoulos et al., 1973). This last result has been justified hypothesizing that the phase transition temperature T_t is near to a natural critical temperature T_c at which the system becomes extremely sensitive to small external perturbations.

Bond and Wyeth show that the isothermal electric susceptibility, χ_T, of a material can be very sensitive to the external electric field at T_c. In fact, they obtain:

$$\chi_T = \frac{1}{\varepsilon_0}\left(\frac{\partial P}{\partial E}\right)_T \propto \frac{1}{a(T)(T-T_c)} \qquad (36)$$

where the parameter a(T) is related to the expansion coefficient in the Landau expansion for free energy (Reichl, 1980). It is clear from this expression that the susceptibility goes to infinite as T goes to T_c. This result is highly qualitative, but can be supported by the consideration that the phosphocholine head group, involved in this phase transition, has been estimated to have an high dipole moment ($\approx$19 D) (Shepherd and Buldt, 1978) and thus can be very sensitive to external fields.

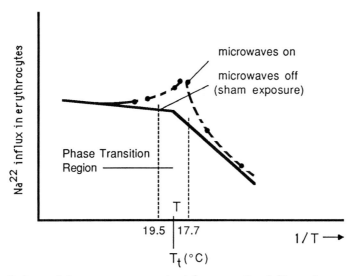

Fig.17. Schematic representation of Liburdy and Penn experimental results.

INTERACTION WITH CHANNEL PROTEINS

Liburdy and Penn (1984), on the basis of the Allis and Sinha results (1981,82), discard the hypothesis that a phase transition can be modified by a microwaves field and suppose that the transmembrane proteins may be the primary target for microwave radiation. The Allis and Sinha experiment fails in confirming the results on a phase transition in multilamellar phospholipid vesicles obtained by Sheridan et al.(1979). It must be noted, however, as Allis and Sinha observe, that the two experiments are different with respect to the sample concentrations, the microwave frequency (1.0 vs. 2.45 GHz), and the measurement techniques (fluorescence vs. Raman spectroscopy).

Various experimental results indicate that microwave fields can act on channel proteins. The Liburdy-Penn experiments show a release into the bath of relatively low molecular weight (28 kilodalton) proteins from cells due to microwave exposure. An analogous result has been recently obtained in experiments on rat olfactory epithelium with microwave (2.45 GHz, CW and 1÷100 Hz AM, 0.5÷18 W/kg) induced shed from the cell membrane of the specific camphor-binding protein (Philippova et al., 1988). In the same experiment, a decrease in the binding process of the ligand to the receptor has been obtained in intact membranes but not in membrane-fraction suspensions; these results confirm the dependence of field induced effects on the biological complexity of the cell. Alekseyev et al. (1980) and Sandblom et al. (1985) show MW-effects on gramicidine A receptors in artificial membranes. Recent results by D'Inzeo et al.(1988) seem to confirm the transmembrane protein receptors and their internal and external neighborhoods as the most plausible sites of interaction.

The MW induced alterations of the acetylcholine (ACh) receptor behavior in myotubes has been examined using a patch-clamp technique (D'Inzeo et al.,1988; Bernardi et al., 1988). This technique (Hille, 1984) allows to measure the current that flows through a membrane channel when a channel opens. With the transmembrane voltage fixed at a given value, the channel mean conductance, mean open time, and mean closed time (see Fig.18 for definitions) can be calculated recording several events before, during, and after microwave exposure. The exposure (9.75 GHz, CW, 50 $\mu W/cm^2$ incident on the petri dish) produced a sharp increase in the channel mean closed time (Fig. 19), while mean conductance and open time remained substantially unchanged. The experimental results indicate a reversible microwave induced desensitization of the ACh receptor (D'Inzeo et al.,1988). Microwave induced decrease of the cholinergic activity in the frontal cortex of rats has also been confirmed *in vivo* (Lai et al., 1988). These effects could be produced by a direct modification in the ACh binding process or by an action of the EM field on the enzymatic reactions that take place near the membrane surface (D'Inzeo et al., 1988). The alteration mechanisms of enzymatic reactions are described in the last section of this chapter. The alteration mechanisms in proteins' behavior or in the ligand-receptor binding are numerous and cover several frequency ranges.

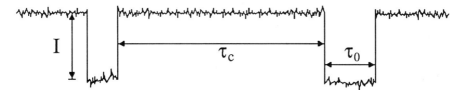

Fig.18. A patch-clamp recording with two openings of a channel. The channel conductance, γ, can be evaluated for each event as the ratio of the current amplitude I to the transmembrane voltage. The open time, τ_O, is the time interval of the current flow, the closed time, τ_C, the time between two events.

The field could alter the affinity between the ligand and the receptor, induce conformational changes in the membrane proteins, release proteins from the membrane, alter the ions' path near the binding site or inside the channel, etc.

An interaction model is proposed by Pickard and Rosenbaum (1978) for the protein gating particle that controls the transmembrane channels and for the ions that surround the part of proteins protruding out of the lipidic substrate. Pickard proposes that the field acts on this gating particle, altering the percentage of membrane ionic channels which are, at any given moment, in a conducting state and, as a consequence, induces macroscopic alterations of the cell behavior.

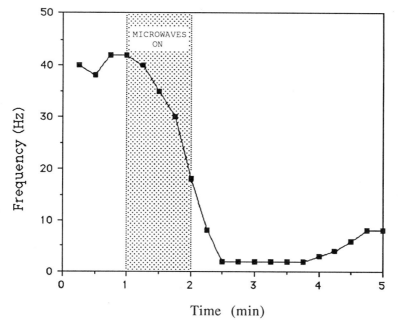

Fig.19. Microwave-induced decrease of openings' frequency (number of events per second) in a chick myotube.

Since the turn-on of a sodium channel in nerve cells is believed to involve the physical displacement of a charged group tethered within the channel, and other channel types could behave similarly, we shall assume that ion gating requires in general the mechanical flipping of a charged particle. A tethered particle of mass m and charge q is held in an equilibrium position by the field (V_0/L) associated with the resting potential V_0 of a membrane of thickness L. As a result of a thermal collision it is suddenly hurled away from the initial position with energy kT/2 and assumes a periodic motion. The characteristic frequency is the reciprocal of its time of flight from rest to apogee and back to rest. By equating the inertial and the restoring forces:

$$m \frac{d^2 x}{dt^2} = - q \frac{V_0}{L} \tag{37}$$

and by solving (37) with the initial conditions:

$$x(0) = 0 \quad ; \quad \left. \frac{dx}{dt} \right|_{t=0} = \sqrt{\frac{kT}{m}} \tag{38}$$

we obtain the characteristic frequency:

$$f_R = \frac{qV_0}{2L} \frac{1}{\sqrt{mkT}} \tag{39}$$

For a tetravalent particle of molecular weight 600, a resting potential $V_0 = 65$ mV, and a membrane thickness $L = 10^{-8}$ m, it follows $f_R \approx 40$ GHz. Gating particle displacement offers a possible resonant mechanism by which microwave irradiation could affect the electrical properties of cells: a change of the percentage of channels in the conducting state alters the membrane's ionic permeabilities, and, consequently, shifts the resting potential.

A different interaction mechanism is proposed by Cain (1980,81) that supposes the applied field to be the cause of a variation of the instantaneous voltage across the membrane. The induced oscillation of the membrane potential could influence the voltage sensing charged groups of the protein macromolecules that form voltage-sensitive ion channels. As a consequence, a modification of the channel opening and closing probabilities is obtained. This variation shifts the mean level of the rate constants that control the conductances of sodium and potassium ionic channels. A dc displacement of the membrane potential from its resting value is the final consequence of this process that is supposed to occur without any change of the total current through the membrane. This nonlinear interaction mechanism produces an inhibition of the neuronal cell activity, in agreement with some experimental effects induced at cellular level by exposure to CW or modulated fields (Lords et al., 1973; Wachtel et al., 1975; Arber, 1976; Tinney et al.,1976; Seaman and Wachtel, 1978; Arber, 1981; Seaman et al., 1982; Arber and Lin, 1983-85; Caddemi et al.,1986; Galvin and Mc Ree, 1986).

INTERACTION WITH IONS AND LIGANDS

Interaction with ions has been proposed as a possible explanation of EM induced flux of calcium observed in vivo and in vitro in different excitable cells (see the complete review made by Adey, 1984). These results include calcium-ions influx or efflux from the cerebral cortex of cats in vivo (Kaczmarek and Adey, 1974) and from isolated chick cerebral hemisphere (Bawin et al. 1975) produced by modulated VHF. Similar subsequent results in the ELF-range (Bawin and Adey, 1976,1977) or with ELF modulated microwaves (Adey et al., 1982) have been confirmed and partially expanded by other authors (Blackman et al., 1979-85; Athey, 1981; Merrit et al., 1982; Dutta et al., 1984). In particular, Blackman et al.(1985) showed that EM-field induced alterations of ion fluxes strongly depend on the value of the static magnetic field; so that two experiments using the same EM-field (amplitude and frequency) may give rise to different results if the environmental static magnetic field is different. This conclusion not only can justify some disparities in the experimental results, but also outlines the importance of the electric and magnetic fields' combined action on exposed biological systems.

A mechanism that can justify Blackman's results has been proposed by Chiabrera et al. (1984, 1985b) by analyzing the EM field local effects on proteins and ions that interact with each other on the surface of the cell. They consider the action of electric and magnetic forces on different ionic messengers and binding sites. The binding process is clearly a statistical event: the binding site moves on the cell surface, while the external messenger (ligand) moves in the microenvironment outside the cell. An applied EM field modifies the velocity and the path of the external messenger near the binding site and will therefore change the association and dissociation rate constants of the receptor-ligand binding action altering the calcium flux through the membrane. The Lagrangian representation of the particle's motion is:

$$d\mathbf{v}/dt + \beta\mathbf{v} + \gamma_B \mathbf{B}_t \times \mathbf{v} = \mathbf{n}_r + \gamma_B \mathbf{E}_{end} + \gamma_B \mathbf{E} \qquad (40)$$

where $\mathbf{v}$ is the ligand velocity, β the ratio of the local viscous coefficient to the ligand mass, γ_B the ratio of the ligand charge to the mass, $\mathbf{E}_{end}$ the local endogenous field, $\mathbf{B}_t$ and $\mathbf{E}$ the local induced fields and $\mathbf{n}_r$ is a random noise with zero average. In Eq. (40) effects of the applied EM field are appreciable only if the interaction takes place in water-free zones so that the field can effectively act on the ligand through the Lorentz and electric forces. Equation (40) is generally difficult to solve analytically and must be solved numerically, but in some particular cases results have been obtained in a closed form both for an harmonic electromagnetic field and for an ELF field superimposed on a static magnetic field (Chiabrera et al., 1985b; Chiabrera and Bianco, 1987). The effects on the velocity of a messenger due to a local EM field of the type $\mathbf{E} = E_1 \sin \omega t \, \mathbf{x}_o$, $\mathbf{B} = (B_0 + B_{ac} \sin \omega t) \mathbf{z}_o$ are shown in Fig.20. The results show the existence of either sensitive or insensitive zones depending on the mass and the charge of the messenger, the frequency and the amplitude of the applied ac field, and the amplitude of the static magnetic field. In particular, remarkable alterations are obtained for integer values of the ratio between the cyclotron resonant frequency $\omega_0 = \gamma_B B_0$ and the angular frequency ω of the ac component. Preliminary results seeming to confirm this theory have been presented (Chiabrera et al., 1986; Pilla et al., 1987). They show the possibility of affecting the swimming direction of a unicellular structure (Paramecium), whose velocity is determined by the flux of calcium through its membrane; the chosen exposure conditions (ac and dc fields) and the frequency allow the modification of the Paramecium behavior in a predictable manner.

An interaction mechanism has been proposed by Liboff and McLeod (1985,1986,1988), in order to justify the results obtained by Blackman et al.(1985). They hypothesize a frequency selective effect based on the cyclotron resonance frequency (CRF) of the ions. An ion moving with a velocity **v** in a static magnetic field **B₀**, covers an helicoidal path with the axis parallel to **B₀**. The motion can be decomposed in a drift displacement at a constant velocity along the **B₀** direction and in a uniform circular movement in the plane normal to **B₀** at the cyclotron resonance frequency. A superimposed magnetic or electric field, oscillating at the CRF frequency, produces a force on the moving ion with components both in the direction of the helicoidal axis and on the plane normal to it.

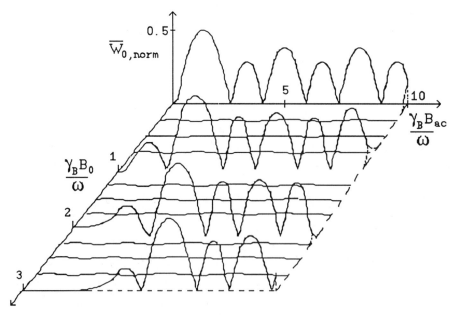

Fig. 20. Normalized time average of the velocity change, $\overline{W}_0$, as a function of the static and ac magnetic field amplitudes. The velocity change is normalized according to the exposure condition (Chiabrera and Bianco, 1987).

By supposing that the ionic movements are guided by helicoidal pathways present in the proteins' structure of the channel walls, or that the quasi-cylindrical structure of the channels screens the ions from the thermal noise present on both sides of the membrane, the ions could be accelerated by the field and cross the membrane channel.

Experimental results showing the strong influence of electric and magnetic combined fields at the CRF condition were given by Thomas et al. (1986) and by Smith et al.(1987). Thomas and coauthors analyzed the behavior of rats under different exposure conditions: sham (no field), static magnetic field, oscillating field, both static and oscillating (at the CRF of Li^+ ions) fields. The rats were tested by evaluating their ability to obtain food; the results showed a modified behavior only in the fourth experiment. The authors relate this result to "the resonant efflux of lithium ions from cells in rat brains". Smith et al.(1987) carried out experiments on diatoms (Amphora coffeaeformis) whose mobility depends on the calcium that crosses the membrane. At a fixed concentration of Ca^{2+} in the bath, the comparison between samples exposed at the CRF condition and sham gives even a tenfold increase in the mobility of the exposed diatoms (Fig.21). Recently Halle (1988), referring to the Liboff and McLeod theory, criticises their proposed interaction mechanism. He affirms that "the cyclotron resonance model is untenable. If the ionic trajectory is prescribed, the magnetic effect vanishes identically. And even if the motional constrains are relaxed" a Langevin approach must be used and, as a consequence, "dynamic friction ensures that the magnetic effect is utterly insignificant". Considering a Langevin approach to ions motion in presence of combined electric and magnetic fields, Durney et al.(1987,1988) apply a numerical technique on a simplified form of equation (40). They conclude

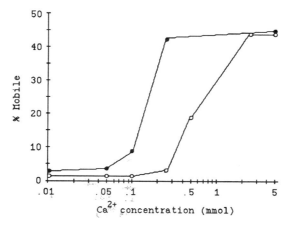

Fig.21. Mobility (in percent of cells that moved more than 10 μm) versus Ca^{2+} concentration. Open dots = control (no field), black dots = experiments tuned for Ca^{2+} resonance condition (CRF =16 Hz at B_0 =20.9 μT, B_{ac} = 20.9 μT). The effect is significant at 0.1, 0.25, and 0.5 mmol, at lower values the effect on calcium was too low, at higher values the phenomenon saturates.

(1988) that "no resonant response is possible unless the viscous damping is very low, many orders of magnitude lower than the viscous damping of ions in solution" and therefore the interaction must take place in "sites other than the motion of bulk ions in solution, perhaps in membrane channel, for example".

INTERACTION AT MEMBRANE SURFACES OR WITH ENZYMATIC REACTIONS

In order to justify the EM field induced effects on the membrane behavior, Blank (1978-1988) concentrates his attention on the ionic processes at the membrane surfaces. At the membrane (or channel) surfaces the surface concentrations and the surface potentials of ions differ from those in the bulk, but the combined electrochemical potentials in the two zones are equal. In the calculation of the ionic behavior around and through the membrane this difference must be considered; in fact the ionic fluxes depend on the electrochemical potential differences, while the ionic permeability is mainly controlled by the surface concentration. This difference is more effective during transient membrane modifications (i.e. action potentials) when the change in the surface charge can take place in microseconds, while chemical adjustments happen with rates typical of the diffusion processes inside the membrane (milliseconds).

Blank (1984, 1988) proposes the surface compartment model (SCM) as "a model of ion transport that focuses on the electrical double layer regions of charged membranes". The SCM model is a simplified version, based on the Helmholtz approach, of the electrical double layer theory applied to the membrane ionic processes. In particular, the membrane environment is divided in five zones: outside, outside surface, membrane, inside surface, inside. In the two surface zones a mechanism of injection and depletion with a spatial dependence linear over the length of the zones is considered for each ion (sodium, potassium, ionic pump). For example, for a sodium ion in the outside surface zone, of thickness L, the following equation is derived:

$$\frac{dN_{Na}}{dt} = \frac{1}{L}\left(J_{Nao} - J_{Nam} - J_{Nap} - \frac{dS_{Na}}{dt}\right) \qquad (41)$$

where N_{Na} is the sodium concentration in the surface zone, J_{Nao} and J_{Nam} are the current-density contributions from the outside zone and to the membrane, J_{Nap} is the current-density contribution due to the sodium pump of the membrane, and S_{Na} is the surface concentration of the sodium controlled by ion binding. The last chemical process is described, in turn, by the equation :

$$\frac{dS_{Na}}{dt} = K^+ N_{Na} X^S - K^- S_{Na} \qquad (42)$$

where K^+ and K^- are the forward and reverse kinetic constants of the reaction, respectively, and X^S is the surface charge concentration at the membrane surface. This charge is related to the sodium and potassium surface concentration rates, and to the gating current density J_G, which arise from changes in the membrane polarization, by:

$$\frac{dX^S}{dt} = -\left(\frac{dS_{Na}}{dt} + \frac{dS_K}{dt}\right) - J_G \qquad (43)$$

Similar equations can be derived in the inside surface zone both for sodium and potassium ions. The law of charge conservation in the five different zones connects all the contributions examined in a unique set of equations. The effects of an external field can be calculated as an injection of charges from the outside to the inside of the membrane for one versus of the field and viceversa for the opposite one. The study of this model, in voltage-clamp conditions, gives for a squid axon membrane a different selection of ion flow through the membrane due to the different kinetics of the channels opening. The consequence is that one period of a sinusoidal signal can give steady-state alterations in the surface ionic concentrations, inducing a shift in the membrane resting potential. The induced alterations in the two sides of the membrane are additive. These alterations are frequency sensitive, since the different dynamic behaviors of the channels, and the ion-concentration variations can modify the channel proteins environment, and change their permeabilities. As a consequence, the process could be auto-catalytic.

A modification of some intracellular chemical reactions (e.g. enzymatic reactions) can greatly modify the cell behavior. This process has been described by Albanese and Bell (1984) examining an enzyme-catalyzed chemical reaction ($X \rightarrow C$) occurring in a fixed volume.

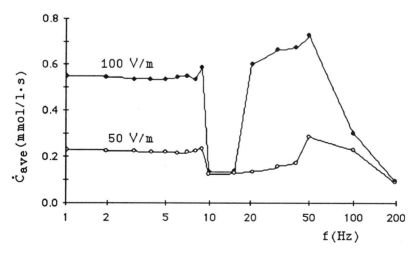

Fig.22. Average of the rate of change of product formation in an enzymatic reaction induced by a modulated microwave field versus the frequency of the modulating signal.

Indicating with k(T) the rate at which the reaction takes place and with X(T,t) the concentration of the reactant, the time rate of change of product formation is:

$$dC/dt = k(T) X(T,t) \qquad (44)$$

Equation (44) together with the nonlinear differential equations that control the enzymatic process simulate the reaction dynamics. The heating produced by a low-frequency modulated microwave signal can indirectly modify the time-rate of the enzymatic reaction by altering the thermodynamics of the reaction. The computed results for an applied field at 400 MHz reveal a slight perturbation of the temperature (ΔT less than 0.01 K) caused by the field, but a significant alteration of the reaction rate constants. The analysis shows a strong dependence of the reaction rate on the amplitude of the carrier field and on the frequency of the modulating signal. For example, in Fig.22 the arithmetic average $\overset{\circ}{C}_{ave}$ of 200 computed values of $\overset{\circ}{C} = k(T)X$ equispaced in time over one period of the modulating frequency is represented as a function of the modulating frequency for different field strengths.

REFERENCES

Adey, W. R., Bawin, S. M., and Lawrence, A. F., 1982, Effects of weak amplitude-modulated microwave fields on calcium efflux from awake cat cerebral cortex, Bioelectromag., 3: 295.

Adey, W. R., 1984, Nonlinear, nonequilibrium aspects of electromagnetic field interactions at cell membranes, in: "Nonlinear Electrodynamics in Biological Systems," W. R. Adey, and A. F. Lawrence, eds., Plenum Press, New York.

Adey, W. R., and Lawrence, A. F., 1984, "Nonlinear Electrodynamics in Biological Systems," Plenum Press, New York.

Albanese, R. A., and Bell, E. L., 1984, Radiofrequency radiation and chemical reaction dynamics, in: "Nonlinear Electrodynamics in Biological Systems," W.R. Adey, and A.F. Lawrence, eds., Plenum Press, New York.

Alekseyev, S. I., Tyazhelov, V. V., Grigor'ev, P. A., and Siden, G. I., 1980, Some aspects of microwave effect on the bilayer membranes modified with gramicidin, Biophysics, 25: 750.

Allis, J. W. , Sinha, B. L., 1981, Fluorescence depolarization studies of red cell membrane fluidity. The effect of exposure to 1.0 GHz microwave radiation, Bioelectromag., 2:13.

Allis, J. W. , Sinha, B. L., 1982, Fluorescence depolarization studies of the phase transition in multilamellar phospholipid vesicles exposed to 1.0 GHz microwave radiation, Bioelectromag., 3:323.

Arber, S. L., 1976, Effect of microwaves on resting potential of giant neurons of mollusk, Helix pomatia, Elektron. Obrab. Mater, 6:78.

Arber, S. L.,1981, The effect of microwave radiation on passive membrane properties of snail neurons, J. Microwave Power, 16:15.

Arber, S. L., and Lin, J. C., 1983, Microwave enhancement of membrane conductance in snail neurons: role of temperature, Physiol. Chem. Phys., 15:259.

Arber, S. L., and Lin, J. C., 1984, Microwave enhancement of membrane conductance in snail neurons: effects of EDTA, caffeine and tetracaine, Physiol. Chem. Phys., 16:469.

Arber, S. L., and Lin, J. C., 1985a, Microwave-induced changes in nerve cells: effects of modulation and temperature, Bioelectromag., 6:257.

Arber, S. L., and Lin, J. C., 1985b, Extracellular calcium and microwave enhancement of membrane conductance in snail neuron, Radiat. Environ., 24:149.

Athey, T. W., 1981, Comparison of RF-induced calcium efflux from chick brain at different frequencies: do the scaled power density windows align?, Bioelectromag., 2:407.

Bawin, S. M., Kaczmarek, K. L., and Adey, W. R., 1975, Effects of modulated VHF fields on the central nervous system, Ann. N.Y. Acad. Sci., 247:74.

Bawin, S. M., and Adey, W. R., 1976, Sensitivity of calcium binding in cerebral tissue to weak environmental electric fields oscillating at low frequencies, Proc. Natl. Acad. Sci. U.S.A., 73:1999.

Bawin, S. M., and Adey, W. R., 1977, Calcium binding in cerebral tissues, in: "Biological Effects and Measurement of Radiofrequency Microwaves," D. G. Hazzard, ed., HEW Publ., FDA.

Bawin, S. M., Sheppard, A. R., and Adey, W. R., 1978, Possible mechanisms of weak electromagnetic field coupling in brain tissue, Bioelectrochem. Bioenerg., 5:67.

Bernardi, P., D'Inzeo, G., Eusebi, F., Grassi, F., and Tamburello, C., Microwave-induced desensitization of acetylcholine receptor channels in cultured quail myotubes, 10th Annual Meet. of Bioelectromag. Soc., (Abstr.), Stamford, USA.

Blank, M., and Britten, J. S., 1978, The surface compartment model of the steady state excitable membrane, Bioelectrochem. Bioenerg., 5:528.

Blank, M., and Kavanaugh W.P., 1982, The surface compartment model (SCM) during transients, Bioelectrochem. Bioenerg., 9:427.

Blank, M., 1983, The surface compartment model (SCM) with a voltage sensitive channel, Bioelectrochem. Bioenerg., 10:451.

Blank, M., 1984, Properties of ion channels inferred from the surface compartment model (SCM), Bioelectrochem. Bioenerg., 13:93.

Blank, M., 1987, Ionic processes at membrane surfaces: the role of electrical double layers in electrically stimulated ion transport, in " Mechanistic Approaches to Interaction of Electromagnetic Fields with Living Systems", M. Blank, and E. Findl, eds., Plenum Press, New York.

Blank, M., 1987, The influence of surface charge on oligomeric reactions as a basis for channel dynamics, in " Mechanistic Approaches to Interaction of Electromagnetic Fields with Living Systems", M. Blank, and E. Findl, eds., Plenum Press, New York.

Blackman, C. F., Elder, J. A., Weil, C. M., Benane, S. G., Eichinger, D. C., and House, D. E., 1979, Induction of calcium ion efflux from brain tissue by radiofrequency radiation: effects of modulation, frequency and field strength, Radio Sci., 14(6S):93.

Blackman, C. F., Benane, S. G., Elder, J. A., House, D. E., Lampe, J. A., and Faulk, J. M., 1980, Induction of calcium ion efflux from brain tissue by radiofrequency radiation: effect of sample number and modulation frequency on the power density window, Bioelectromag., 1:35.

Blackman, C. F., Benane, S. G., Rabinowitz, J. R., House, D. E., and Joines, W. T., 1985, A role for the magnetic field in the radiation-induced efflux of calcium ions from brain tissue in vitro, Bioelectromag.,6:327.

Bond, J. D., and Wyeth, N. C., 1986, Are membrane microwave effects related to a critical phase transition?, J. Chem. Phys., 85(12):7377.

Bond, J. D., and Wyeth, N. C., 1987, Membrane, electromagnetic fields, and critical phenomena, in: "Mechanistic Approaches to Interaction of Electromagnetic Fields with Living Systems", M. Blank, and E. Findl, eds., Plenum Press, New York.

Caddemi, A., Tamburello, C., Zanforlin, L., and Torregrossa, V., 1986, Microwave effects on isolated chick embryo hearts, Bioelectromag., 7:359.

Cain, C.A., 1980, A theoretical basis for microwave and RF field effects on excitable cellular membranes, IEEE Trans. MTT, 28:142.

Cain, C.A., 1981, Biological effects of oscillating electric fields: role of voltage sensitive ion channels, Bioelectromag., 2:23.

Cevc, G., Marsh, D., 1987, "Phospholipid bilayers: physical principles and models,", John Wiley & Sons, New York.

Chiabrera, A., Grattarola, M., and Viviani, R., 1984, Interaction between electromagnetic fields and cells: Microelectrophoretic effect on ligands and surface receptors, Bioelectromag., 5:173.

Chiabrera, A., Nicolini, C., and Schwan, H.P., 1985a, "Interaction Between Electromagnetic Fields and Cells," Plenum Press, New York.

Chiabrera, A., Bianco, B., Caratozzolo, F., Giannetti, G., Grattarola, M., and Viviani, R., 1985b, Electric and magnetic field effects on ligand binding to the cell membrane, in:"Interaction between Electromagnetic Fields and Cells", A. Chiabrera, C. Nicolini, and H.P. Schwan, eds., Plenum Press, New York.

Chiabrera, A., Gyebyi, K., Kaufman, J., Ryaby, J., Smith Sonneborn, J., and Pilla, A.A., 1986, Lorentz magnetic force effect on biological systems: application to paramecium, 6th Annual Meet. of BRAGS, (Abstr.), Utrecht, The Netherlands.

Chiabrera, A., and Bianco, B., 1987, The role of the magnetic field in the em interaction with ligand binding, in " Mechanistic Approaches to Interaction of Electromagnetic Fields with Living Systems", M. Blank, and E. Findl, eds., Plenum Press, New York.

Collin, R. E., 1966, "Foundations for Microwave Engineering", Mc Graw-Hill, New York.

D'Inzeo, G., Bernardi, P., Eusebi, F., Grassi, F., Tamburello, C., and Zani, B.M, 1988, Effects of microwaves on the acetylcholine-induced channels in cultured chick myotubes, Bioelectromag., 9:363.

Durney, C. H., Rushforth, G. K., and Anderson, A. A., 1988, Resonant AC-DC magnetic fields: calculated response, Bioelectromag., 9:315.

Durney, C. H., Anderson, A. A., and Rushforth, G. K., 1987, Calculated biological response to resonant DC-AC magnetic fields, 9th Annual Meet. of Bioelectromag. Soc., (Abstr.), Portland, USA.

Dutta, S. K., Subramoniam, A., Ghosh, B., and Parshad, R., 1984, Microwave radiation-induced calcium ion efflux from human neuroblastoma cells in culture, Bioelectromag., 5:71.

Foster, K. R., and Schwan, H. P., 1986, Dielectric properties of tissues, in: "CRC Handbook of Biological Effects of Electromagnetic Fields", C. Polk, and E. Postow, eds., CRC Press, Boca Raton.

Fröhlich, H., 1950, "Theory of Dielectrics: Dielectric Constant and Dielectric Loss", Oxford University Press, London.

Galvin, M. J., and McRee, D. I., 1986, Cardiovascular, hematologic and biochemical effects of acute ventral exposure of conscious rats to 2450-MHz (CW) microwave radiation, Bioelectromag., 7:223.

Genzel, L., Kremer, F., Poglitsch, A., and Bechtold, G., 1983, Relaxation processes on a picosecond time scale in hemoglobin and poly (L-Alanine) observed by millimeter-wave spectroscopy, Biopolymers, 22:1715.

Halle, B., 1988, On the cyclotron resonance mechanism for magnetic field effects on transmembrane ion conductivity, Bioelectromag., 9:381.

Hille, B., 1984, "Ionic Channels of Excitable Membranes;" Sinauer Associates Inc., Sunderland.

Kaczmarek, L. K., and Adey, W. R., 1974, Weak electric gradients change ionic and transmitter fluxes in cortex, Brain. Res., 66:537.

Kremer, F., Poglitsh, A., and Genzel, L., 1984, Picosecond relaxations in proteins and biopolymers observed by MM-wave spectroscopy, in :"Nonlinear Electrodynamics in Biological Systems," W. R. Adey, and A. F. Lawrence, eds., Plenum Press, New York.

Lai, H., Horita, A., and Guy, A. W., 1988, Acute low-level microwave exposure and central cholinergic activity: studies on irradiation parameters,Bioelectromag., 9:355.

Liboff, A. R., 1985, Cyclotron resonance in membrane transport, in: "Interaction Between Electromagnetic Fields and Cells", A. Chiabrera, C. Nicolini, and H. P. Schwan, eds., Plenum Press, New York.

Liboff, A. R., and McLeod, B. R., 1988, Kinetics of channelized membrane ions in magnetic fields, Bioelectromag., 9:39.

Liburdy, R. P., Penn, A., 1984, Microwave bioeffects in the erytrocyte are temperature and pO dependent: cation permeability an protein shedding occur at the membrane phase transition, Bioelectromag., 5:283.

Liburdy, R. P., 1985, Evidence that microwave stimulate free radical production: autooxidation of hemoglobin to produce superoxide, <u>7th Annual Meet. of Bioelectromag. Soc.</u>, (Abstr.), San Francisco, USA.

Liburdy, R. P., 1986, Microwave associated protein shedding in human herythrocite: quantitative HPLC analysis, <u>8th Annual Meet. of Bioelectromag. Soc.</u>, (Abstr.), Madison, USA.

Lin, J. C., 1989, " Electromagnetic Interaction with Biological System," Plenum Press, New York.

Lords, J. L., Durney, C. H., Borg, A. M., and Tinney, C. E., 1973, Rate effects in isolated hearts induced by microwave irradiation, <u>IEEE Trans. MTT</u>, 21:834.

McLeod, B. R., and Liboff, A. R., 1986, Dynamic characteristics of membrane ions in multifield configurations of low frequency electromagnetic radiation, <u>Bioelectromag.</u>, 7:177.

Merritt, J. G., Shelton, W. S., and Chamnes, A. F., 1982, Attempt to alter $^{45}Ca^{2+}$ binding to brain tissue with pulse-modulated microwave energy, <u>Bioelectromag.</u>, 3:475.

Michaelson, S. M., and Lin, J. C., 1987, " Biological Effects and Health Implications of Radiofrequency Radiation," Plenum Press, New York.

Milazzo, G., 1985, Bioelectrochemistry and Bioelectromagnetism, <u>Bioelectrochem. and Bioenerg.</u>, 14:5.

Papahadjopoulos, D., Jacobson, K., Nir, S., and Isac, T., 1973, Phase transitions in phospholipid vesicles. Flourescence polarization and permeability measurements concerning the effect of temperature and cholesterol, <u>Biochim. Biophys. Acta</u>, 311:330.

Pethig, R., 1979, "Dielectric and Electronic Properties of Biological Materials," John Wiley & Sons, Chichester.

Philippova, T. M., Novoselov, V. I., Bystrova, M. F., and Alekseev, S. I., 1988, Microwave effects on champhor binding to rat olfactory epitelium, <u>Bioelectromag.</u>, 9:347.

Pickard, W. F., and Rosenbaum, F. J., 1978, Biological effects of microwaves at the membrane level: two possible athermal electrophysiological mechanisms and a proposed experimental test, <u>Math. Biosci.</u>, 39:235.

Pilla, A. A., Chiabrera, A., Kaufman, J. J., and Ryaby J. T., 1987, A unified electrochemical approach to electrical and magnetic modulation of biological processes: application to the paramecium ciliary movement, <u>9th Annual Meet. of Bioelectromag. Soc.</u>, (Abstr.) F-6, Portland, USA.

Polk, C., and Postow, E., 1986, "Handbook of Biological Effects of Electromagnetic Fields," CRC Press, Boca Raton, California.

Reichl, L. E., 1980, "A Modern Course in Statistical Physics," University of Texas, Austin.

Sandblom, J., Teander, S., and Baltzer, P., 1985, The effect of microwave radiation on the stability and formation of gramicidin A channels in lipid bilayer membranes, <u>Proc. of XIV ICMBE</u>, Espoo, Finland.

Scherer, P.G., and Seelig, J, 1987, Reorientation of lipid head groups at membrane-water interface, <u>EMBO J.</u>, 6:2915.

Seaman, R., and Wachtel, H., 1978, Slow and rapid response to CW and pulsed microwave radiation by individual Aplysia pacemakers, J. Microwave Power, 13:77.

Seaman, R., Ajer, R. K., and DeHaan, R. L., 1982, Changes in cardiac-cell membrane noise during microwave exposure, Proc. of the IEEE-MTT Symp., 1:436.

Seelig, J., Macdonald, P.M., and Scherer, P.G., 1987, Phospholipid head groups as sensors of electric charge in membranes, Biochem., 24:7535.

Shepherd, J. C. W., and Büldt, G., 1978, Dielectric measurements on reorientation rates in lipid head groups, Biochim. Biophys. Acta, 514:83.

Sheridan, J. P., Gaber, B. P., Cavatorta, F, and Schoen, P. E., 1979, Molecular level effects of microwaves on natural and model membranes: a Raman spectroscopic investigation, in, "Natl. Radio Science Meet. and Bioelectromag. Symp.", L.S. Taylor, and A. Y. Cheung, eds, Seattle, University of Washington.

Smith, S. D., McLeod, B. R., Liboff, A. R., and Cooksey, K., 1987, Calcium cyclotron resonance and diatom mobility, Bioelectromag., 8:218.

Thomas, J. R., Schrot, J., and Liboff, A. R., 1987, Low-intensity magnetic fields alter operant behavior in rats, Bioelectromag., 7:349.

Tinney, C. E., Lords, J.L., and Durney, C. H., 1976, Rate effects in isolated turtle hearts induced by microwave irradiation, IEEE Trans. MTT, 24:18.

von Hippel, A. R., 1954, "Dielectrics and Waves", J. Wiley & Sons, New York.

Wachtel, H., Seaman, R., and Joines, W., 1975, Effects of low-intensity microwaves on isolated neurons, Ann. N.Y. Acad. Sci., 247:46.

BIOLOGICAL EFFECTS OF RADIOFREQUENCY

RADIATION: AN OVERVIEW

Stephen F. Cleary

Department of Physiology
Medical College of Virginia, Virginia Commonwealth
University, Richmond, Virginia 23298 U S A

INTRODUCTION

The purpose of this chapter is to provide a biological perspective for radiofrequency radiation safety standards. Rather than a detailed consideration of specific safety standards now in effect, or proposed safety standards, general problems involved in integrating available bioeffects data into practical and effective safety standards are discussed. The reader will find other aspects of standard setting and enforcement discussed elsewhere in this volume. Engineering problems related to RF-field coupling to the human body and the measurement and characterization of incident and internally induced RF fields, as well as cost-benefit considerations, unquestionably play a significant role in the development of standards. The approach taken here, of focussing primarily on the biological basis of RF standards, should not be construed as an attempt at diminishing the significance or inadequacy of these other aspects of standard setting. However, in the final analysis the most important aspect of a safety standard must be its adequacy in preventing adverse health effects in occupationally or environmentally exposed humans. This cannot be assured in the absence of an adequate understanding of the biological effects of RF radiation. Unfortunately, in spite of well over two decades of research, the adequacy of our knowledge of RF bioeffects is questionable. The remainder of this chapter is devoted to a discussion of some inadequacies in the RF bioeffects data base and how and why they evolved.

Historically, reports of the biological effects of RF and lower frequency electromagnetic radiation date back to workers such as d'Arsonval at the turn of this century. Based upon the research of early workers in this field, the hypothesis gradually emerged that the effects of RF radiation were indirect, nonspecific, thermal effects. Unquestionably, absorption of RF radiation at a rate exceeding the body's capacity to redistribute or dissipate the added energy burden causes tissue temperature elevations which if intense and/or persistent enough will result in thermal damage. Early safety standards, not surprisingly, were based upon limitation of RF-induced body core-temperature elevations to a maximum of 1°C, a thermal burden of the same approximate magnitude as experienced during febrile states.

This approach to setting safety standards, based upon general or nonspecific thermal physiological responses was logical at the time. However, evidence began to accumulate from RF bioeffects studies, in the late 50's and early 60's, that it was overly simplified. Indeed, there are many examples of comparisons of RF thermal effects with effects due to other modalities of thermal stress, such as elevated environmental temperature, that revealed significant qualitative and quantitative differences in response in situations that involved equivalent core or rectal temperature elevations in laboratory animals. The concept thus emerged that instead of nonspecific thermal effects, RF radiation could also induce RF-specific thermal effects (Cleary, 1977). Subsequent attempts to explain RF-specific thermal effects required advances in RF dosimetry and densitometry that occurred primarily in the 70's and early 80's. Experimental and theoretical techniques were developed which revealed that RF radiation absorption in vivo was in general a highly nonuniform wavelength-dependent phenomenon. This knowledge provided a basis for explaining the existence of RF-specific thermal effects since it was now obvious that RF and environmental heat stress were not directly comparable; thus, they could not be expected to result in the same type or extent of physiological alterations. Recognition of RF-specific thermal bioeffects advanced the general state of knowledge in this field. However, awareness of such phenomena introduced significant difficulties in the interpretation of RF bioeffects data involved, for example, with interspecies extrapolation of bioeffects data from laboratory animals to humans, as discussed below. Since there is a very limited RF bioeffects database for human exposures, RF safety standards have relied almost exclusively on animal experimental data. The unavoidable expediency of interspecies extrapolation of data has presented significant problems in the setting of RF safety standards.

INTERSPECIES EXTRAPOLATION

In addition to complexities introduced into setting of safe exposure standards due to the induction of RF-specific thermal effects in living systems, a much more fundamental problem exists, namely the lack of understanding of interaction mechanisms responsible for RF alterations. Even if it were assumed that all RF effects on living systems were indirect thermal effects, albeit RF-specific effects due to nonuniform energy deposition, mechanistic understanding, and hence predictive ability, would still be lacking due to several factors. Principally, uncertain or incomplete knowledge of thermal physiology in the various species of laboratory animals used in RF-bioeffects studies would make it difficult to predict effects of a novel type of thermal stress in a given species exposed to RF radiation, much less to extrapolate between species as required in utilizing bioeffects data for the setting of safety standards. (Cleary, 1983a).

RF SPECIFIC THERMAL BIOEFFECTS

As an example of this problem, consider the results of D'Andrea and coworkers (1985). These workers detected localized regions of RF field concentrations (i.e. electromagnetic "hot spots") in the brain, rectum, and tail of rat carcasses exposed to 2450- or 360 MHz RF radiation. The localized specific absorption rate (SAR) in the tails of rat carcasses exposed to 2450 MHz radiation exceeded the whole-body-averaged SAR by a factor of 18. Exposure to 360 MHz radiation, on the other hand, caused a 50 times greater SAR in the tail, relative to the whole-body-averaged SAR. These results in dead animals provided no information on the effect of blood circulation on limiting the thermal consequences of electromagnetic hot spots in the tail. Consequently, D'Andrea et al.

(1987) conducted a follow-on study using anesthetized rats of the same strain and size as in the initial study to measure localized RF-induced temperature elevations (i.e. "thermal" hot spots). Exposures of 10 to 16 min duration at 2450-, 700- or 360 MHz, at SARs of 2-,6-, or 10 W/kg, caused immediate postexposure temperature elevations at sites previously shown to be electromagnetic hot spots in rat cadavers. Rectal and tail temperatures were significantly higher in anesthetized rats exposed to 360 MHz RF radiation, relative to other frequencies. Tail temperatures were higher in rats exposed to 2450 MHz radiation relative to exposure at 700 MHz. Brain temperatures were also elevated, but to a lesser magnitude than rectal or tail temperatures.

D'Andrea et al. (1987) noted that at an ambient temperature of 24°C, which is below thermoneutrality for the rat, the tail temperature was approximately 24°C, whereas the rectal temperature was 37.5°C; a consequence of normal thermal physiological regulation by this species. Since tail vasodilation is a primary means of thermoregulation in the rat, localized RF heating of this region could have profound effects on the ability of this species to thermoregulate, compared to other species that use different thermoregulatory mechanisms. The data of D'Andrea et al. (1985, 1987) showing frequency dependent heating in the rat tail provide an example of a species-specific RF-specific thermal effect. How could these data be extrapolated to effects on a tail-less animal exposed to the same frequency (e.g. 360 MHz) RF radiation, or to another frequency of RF radiation?

DOSIMETRY AND DENSITOMETRY

Induction of electromagnetic and thermal hot spots in laboratory animals thus provides an experimental basis for RF-specific thermal bioeffects. These phenomena must obviously be taken into account in setting exposure standards. Awareness of such effects has led to attempts to develop theoretical and experimental dosimetric methods that account for frequency dependent field coupling and nonuniform internal absorption, thus providing means of correlating and extrapolating bioeffects data for safety standards.

It was recognized that the coupling efficiency of RF fields to the bodies of laboratory animals or humans, as well as internal distributions of absorbed energy, were dependent upon size, shape, orientation, and dielectric properties of tissue. Coupling of RF fields to animal bodies, for example, was investigated by Gandhi (1975) using antenna theory. The animal's body was modelled as a lossy dielectric antenna with wavelength-dependent variation in coupling to an incident RF field. Gandhi found that the absorption efficiency of a 1.75-m tall erect human weighing 75 kg was maximum at 68 MHz (4.4 m wavelength) in free space. By comparison, this "whole body" resonant condition for a 5.4 cm long mouse weighing 15 g occurred at a frequency of 2.5 GHz (wavelength 12 cm). In addition to wavelength dependence, it was also shown that the amount of coupling depended upon the orientation of the major body axis (L) with respect to RF field polarity. Of three possible orientations: L parallel to the electric field vector (E), the magnetic field vector (H) or the propagation vector (k), maximum coupling occurred for the L parallel to E orientation (Gandhi, 1975).

Awareness of the frequency dependence of whole body RF field coupling and internal distribution of RF energy dictated the need for a means of tissue dose normalization to facilitate the extrapolation and correlation of RF bioeffects data. The concept of a mass and time-averaged rate of RF energy absorption, referred to as the specific absorption rate (SAR), was thus introduced and theoretical and experimental methods of SAR determination were developed. Conceptually, localized SAR's throughout an animal's body could be determined to any degree of resolution, but practically SAR determinations have been generally restricted to whole body averages determined experimentally by calorimetry or theoretically using prolate spheroidal models (Cleary, 1988). In the absence of localized SAR determinations it is obvious that the utility of bioeffects data is compromised, as exemplified by consideration of the data of D'Adrea et al. (1985, 1987) referred to above. If these workers had not measured local SARs in, for example, the rat tail, erroneous conclusions could have resulted regarding relative physiological responses to the RF frequencies they studied. Erroneous predictions of effects in other species could also have occurred using data that did not take localized thermal or electromagnetic hot spots into account.

The limited availability of sophisticated nonperturbing temperature measuring instrumentation and associated methodologies has restricted the number of RF bioeffect studies in which detailed tissue localized SARs have been determined. Thus, in many instances when localized SARs were not determined there is an uncertain correlation between RF exposure parameters and physiological responses. The most obvious way to eliminate such uncertainty would be to determine SAR distributions and/or temperature distributions in vivo. Unfortunately, this has proven difficult, in a practical sense, especially for live, unrestrained laboratory animals. Recent experimental and theoretical advances, however, suggest that such problems may be overcome in future studies. The availability of detailed data on the frequency dependent dielectric properties of live mammalian tissue, sophisticated mathematical models, and high speed digital computers, afford a means of providing theoretical estimates of the intensity and location of hot spots _in vivo_. Armed with this information, an investigator could precisely determine nonuniformities in RF energy absorption experimentally and relate them to physiological end points.

Such advances are a _necessary_ first step toward achieving a more adequate data base for setting RF safety standards. Obtaining a _sufficient_ data base must, however, involve overcoming the other formidable obstacle noted above, namely the absence of knowledge of the mechanisms whereby RF radiation alters living systems, other than by heating. The magnitude of this problem is suggested by the fact that even if it were assumed that _all_ physiological effects of RF radiation were thermally induced, there is at present an incomplete mechanistic understanding of the effects of heat on tissue. It should be noted that there is a growing body of data indicating direct, nonthermal effects of RF radiation on living systems, as discussed below.

MECHANISMS OF RF EFFECTS

Initial attempts to describe the basic mechanisms whereby RF radiation interacts with biomolecules leading to altered function, other than by thermal denaturation, relied on classical approaches. The basic molecular mechanism of interaction of higher frequency electromagnetic radiation, namely ionization leading to disruption of chemical bonds, could not be applied to RF radiation since, in fact, quantized RF energy is significantly less than that required for ionization. Comparison of the magnitude of quantized energy of RF radiation with activation energies for other biomolecular processes, such as hydrogen bond disruption, revealed that even for this class of nonionizing effects, RF quantum energies were still too small to produce alterations (Cleary,

1973). Thus, from the point of view of bioenergetics, nonthermal RF interactions leading to functional alterations in biomolecules would involve resonant absorption of RF radiation since nonresonant interactions have insufficient energy for molecular perturbations. The only known interaction of biomolecules with electromagnetic radiation in the RF frequency range is via field-induced molecular dipole rotation, as described by Debye dispersion theory. Such interactions, dipolar relaxations, are dispersive rather than resonance phenomena, which do not possess the frequency specificity necessary to account for nonthermal RF bioeffects. Thus, there is at present no accepted theoretical basis for low-intensity or nonthermal RF bioeffects.

In light of experimental observations of nonthermal electromagnetic field bioeffects, to be discussed below, the lack of theoretical explanations may appear perplexing. However, it should be noted that attempts to develop theoretical interaction mechanisms have generally not taken the nonequilibrium properties of living systems into account. A vast array of possible nonlinear processes that are known to be involved in living systems have not been incorporated into models of RF bioeffects, except in general, descriptive approaches that invoke concepts such as cooperative interactions. Although cooperative phenomena may eventually provide a theoretical basis for nonthermal RF bioeffects they have not yet provided specific testable hypotheses leading to mechanistic understanding. Indeed, the inability to provide a theoretical basis for nonthermal RF bioeffects may well be attributed in large measure to limited information on the intrinsic cooperative properties of molecular, macromolecular, and molecular assemblages comprising living systems.

In spite of the fact that interaction mechanisms are largely unknown it is of central importance to consider evidence of nonthermal RF bioeffects in the context of this book since Western RF safety standards have generally been based upon the premise that all significant bioeffects are a consequence of heating. What follows is a review of reported RF bioeffects that provide direct evidence of nonthermal mechanisms of interactions. For the purpose of providing a wider perspective, nonthermal effects of electromagnetic fields of frequencies lower than RF radiation will also be summarized briefly. Taken together with RF bioeffects data, these data indicate that living systems exhibit a here-to-fore unexpected sensitivity to electromagnetic fields.

NONTHERMAL ELECTROMAGNETIC FIELD EFFECTS

In view of the highly complex nature of the physical interaction of RF fields with laboratory animals or humans, leading to frequency-dependent field coupling and nonuniform absorbed energy distributions in tissue, it is extremely difficult to conduct *in vivo* experiments under condition of precise or accurate densitometry and dosimetry. Added complexity introduced by the highly interactive nature of mammalian organ systems has made the detection or characterization of direct nonthermal RF bioeffects *in vivo* largely intractable. Consequently, most of the evidence for such effect has been obtained *in vitro*, principally in studies of mammalian cells. One approach involves exposure of cells *in vitro* to RF intensities that are so low that passive heat dissipation mechanisms act to limit induced temperature elevations to the normal range of physiological variability (i.e. approximately $\pm 1°C$). If it is assumed that RF bioeffects have dose or dose rate thresholds, this approach will have a low probability of detecting nonthermal effects due to restricted intensity range.

Alternatively, cells can be exposed to a significantly wider RF intensity range, under isothermal conditions, by utilizing forced convective cooling to actively remove RF-induced heat. Systems for isothermal in vitro cell exposure to RF radiation in the frequency range of 10 MHz to 4 GHz have been described by Cleary et al. (1985a,b) and Liu and Cleary (1988), for example. Detailed experimental verification of isothermal cell exposure conditions is required to insure against artifactual results due, for example, to spatial or temporal temperature gradients, especially at SARs that are of sufficient magnitude to induce significant intrasample temperature elevations in the absence of active temperature control (Cleary et al., 1985a,b; Liu and Cleary, 1988).

ERYTHROCYTES

Availability and the relative structural and functional simplicity of mammalian erythrocytes have made this cell the system of choice for numerous cell biological investigations of phenomena such as membrane transport and energy metabolism. Effects of RF radiation on these cellular endpoints have provided the most extensive body of information indicating nonthermal alterations in vitro.

Since the permeability of cell biomembranes is temperature-dependent, rigid temperature control must be exercised in RF exposure studies. Liu et al. (1979) investigated the effects of continuous wave (CW) 2.45- 3.0- and 3.95 GHz RF radiation on K^+ and Na^+ permeability of rabbit, canine and human erythrocytes, under conditions of precise temperature control. RF intensity was varied from 0 to 200 W/kg to expose red cells at various temperatures from 26 to 44°C. Exposure resulted in RF intensity-dependent alterations in K^+ efflux, Na^+ influx, hemoglobin efflux, and osmotic lysis. However, similar effects were noted in cells simultaneously exposed for the same duration, at the same temperature, in a hot water bath. This led to the conclusion that under these exposure conditions, erythrocyte permeability was not directly affected by RF radiation. Subsequent studies of RF effects on erythrocyte plasmalemma cation permeability, and other endpoints, under different conditions, revealed direct RF effects, however.

Olcerst et al. (1980), for example, exposed rabbit red cells to various intensities of 2.45 GHz CW RF radiation at controlled temperatures in the 4 to 38°C range. Exposure at 100-, 190-, or 390 W/kg resulted in statistically significant effects on passive rubidium (Rb^+) and sodium (Na^+) efflux, at temperatures of 8-,11-, 22.5 and 36°C, but not at other temperatures. Based on studies of the effects of temperature on erythrocyte membrane cation permeability, Cleary et al. (1982) suggested that temperature-specific RF effects may involve membrane phase changes. Although it was known that membrane cation permeability was increased at specific temperatures associated with gel to liquid crystal transitions, it had not been demonstrated previously that RF radiation, or other agents, could enhance cation transport under these conditions. Since RF-enhanced cation fluxes occurred under isothermal conditions, this effect could not be due to RF heating per se.

Subsequent studies provided additional evidence of direct effects of RF radiation on erythrocyte membranes. Cleary et al. (1982) reported increased K^+ release from rabbit erythrocytes exposed to 8.42 GHz RF radiation. It was concluded that this effect was not due to RF induced heating. No difference in K^+ efflux was detected in cells RF- or sham-exposed to various temperature-SAR combinations in the ranges 20- to 30°C and 21- to 90 W/kg, respectively, except at the combination of

24.6°C and 22 W/kg. The magnitude of the RF-induced permeability alteration under this exposure condition was dependent upon physiological factors as well as RF field modulation; pulse-modulation caused consistently greater release them CW radiation. Temperature specificity of the RF-induced enhanced K^+ permeability was consistent with a membrane phase change, as in the study of Olcerst et al. (1980). Differences in the temperatures at which RF enhanced cation permeability occurred in these studies may be attributed to different experimental conditions, including RF frequency.

Additional confirmation of the direct effects of 2.45 GHz CW RF radiation on erythrocyte cation permeability was provided by Liburdy and Penn (1984) and Liburdy and Vanek (1985). In the temperature range 17.7- to 19.5°C these investigators detected RF dose-dependent increased Na^+ permeability, with an apparent electric field-strength threshold of approximately 150 V/m, in general agreement with the field strengths reported by Olcerst et al. (1980) of 353- to 700 V/m and Cleary et al. (1982), 100 V/m. Liburdy and Penn (1984) and Liburdy and Vanek (1985) reported that RF exposure also induced shedding of membrane proteins. Cation release was also dependent upon other factors such as oxygen tension, cholesterol, and antioxidants. It was thus suggested that RF effects on biomembranes were dependent on factors related to the cellular environment and that effects may be more general in nature, rather than restricted to specific transport processes.

Evidence of the importance of the extracellular environment in RF-induced cell alterations was reported by Cleary et al. (1982). RF radiation induced significantly greater alterations in cation fluxes, as well as hemoglobin release, from erythrocytes exposed in whole blood compared to the same cells exposed in isotonic buffered saline after removal of other blood cell types and plasma by centrifugation and cell washing. Similar effects were reported by Liburdy and Penn (1984) and Liburdy and Vanek (1985). Cleary et al. (1979) had previously determined that pulsed electric fields also induced greater effects on plasmalemma permeability of erythrocytes exposed in the presence of autologous plasma, relative to exposure in isotonic buffered saline. Detailed mechanisms for the enhancement of the effects of electromagnetic fields on red cell membrane permeability have not been determined. Cleary et al. (1982) suggested that RF effects do not appear to depend on the presence of leukocytes of serum electrolytes. Plasma components, such as lipoproteins, may be involved in the enhanced cation fluxes in RF exposed red cells at phase change temperatures. It is worth noting that these results suggest that red cells have maximum sensitivity when they are in a normal physiological extracellular environment as would be the case for _in vivo_ exposure. The dependence on such factors indicates the need to conduct RF bioeffects studies _in vitro_ under conditions that simulate _in vivo_ conditions. This does not appear to be the case in most _in vitro_ studies of RF bioeffects.

Studies of the effects of RF radiation on erythrocytes _in vitro_ thus provide firm evidence of direct membrane interactions resulting in permeability alterations and other effects related to the cell environment. A possible mechanism for one such effect, namely alteration in active transport of cations by erythrocyte membranes, was reported by Allis and Sinha-Robinson (1987). Human erythrocyte membrane fragments were exposed to 2.45 GHz CW microwaves at 6 W/kg (approximate induced field strength 90 V/m) _in vitro_. Temperature of the membrane suspension was varied in 1°C increments in the range 23 to 27°C and the activity of membrane Na^+/K^+ ATPase was determined spectrophotometrically. There was a 35% inhibition of ATPase activity at 25°C, but not at other temperatures. Thus, the combination of RF exposure and a specific temperature (25°C) associated with a membrane phase change, altered enzyme activity. Allis and Sinha-Robinson (1987) also detected marginal changes in ouabain-insensitive Ca^{+2} ATPase activity. The involvement of RF-induced heating in enzyme activity alteration was ruled out since Na^+/K^+ ATPase (and Ca^{+2} ATPase) activity would have increased rather than decreased with increased temperature. The results of this study are in general agreement with the results of studies of RF effects on intact red cell cation permeability. Fisher et al. (1982), for example, detected a 40% decrease in ouabain-sensitive Na^+ efflux from human erythrocytes at 23 and 24°C during exposure to 2.45 GHz RF radiation at 3 W/kg. The temperature-specific, RF-induced, alterations in Na^+/K^+ ATPase reported by Allis and Sinha Robinson (1987) are consistent with RF-induced alterations in red cell cation transport reported by Cleary et al. (1982), Olcerst et al. (1980), Liburdy and Penn (1984), and Liburdy and Vanek (1985), discussed above. Since both active and passive red cell membrane transport was altered by RF exposure in the membrane phase change temperature region, it is suggested that this type of radiation may affect more than one membrane transport process. Differences in the temperature or temperature range reportedly involved in RF-induced membrane transport processes can be attributed to differences in temperature dependence of passive versus active transport, as well as known inter- and intra-species variations in membrane composition that affect various membrane transport functions (Claret et al., 1978; Roelofsen, 1981). Membrane variability may also account for reported differences in RF intensities associated with altered cation permeability. It should be noted that with a few exceptions RF dose-response relationships and intensity thresholds have not been reported for RF-induced alteration in cell membrane permeability or other cellular endpoints.

The results of Allis and Sinha-Robinson (1987) provide evidence of a specific molecular level interaction of RF radiation. They concluded that the change in Na^+/K^+ ATPase activity at 25°C resulted from either an alteration in protein-lipid interactions in the membrane or a direct effect of RF radiation on the enzyme. Thus, 2.45 GHz RF radiation may interact with Na^+/K^+ ATPase or with the enzyme-lipid complex in a temperature dependent transition state, resulting in decreased enzymatic activity and hence reduced transmembrane K^+ and Na^+ transport.

Direct nonthermal biomolecular interactions of RF radiation, leading to functional alterations, are thus suggested by studies of erythrocyte cation transport in vitro. Mechanistic considerations dictate the need to consider the frequency specificity of such effects. Due to limited data, however, it is difficult to evaluate the role of radiation frequency in altered erythrocyte cation permeability. The majority of such effects have occurred at a RF frequency of 2.45 GHz. However, Cleary et al. (1982), reported altered K^+ permeability in red cells exposed to CW or pulse-modulated 8.42 GHz RF radiation, suggesting the possibility that this effect may not be frequency-specific. There is evidence that lower frequency RF radiation may induce qualitatively and quantitatively different effects on erythrocytes.

In one such study, Cleary et al. (1985a) detected RF field-strength dependent erythrocyte hemolysis. Hemolysis occurred in suspensions of rabbit blood exposed to 10- 50- or 100 MHz CW RF radiation for 2 h. At 100 MHz the hemolysis field strength threshold was 400 V/m, whereas fields of 900 V/m were required for hemolysis by 10 MHz CW RF radiation. Since temperature was 22.5 ± 0.2°C, the hemolytic effect was not due to RF-induced heating per se. At field strengths that induced hemolysis, there were no detectable alterations in erythrocyte K^+ or Na^+ fluxes, in distinct contrast to the results of previous experiments (Cleary et al.

1982) conducted under similar conditions but at a frequency of 8.42 GHz. These results suggest frequency-dependent differences in RF bioeffects on mammalian erythrocytes. However, there is no evidence (positive or negative) of RF frequency-specific cellular effects.

The effects of lower frequency RF radiation on erythrocytes were reported by Serpersu and Tsong (1983), who investigated effects on Rb^+ uptake in human erythrocytes in vitro. Increased uptake was at a maximum at a field strength of 2 KV/m, at a RF frequency of 1 kHz. Rubidium uptake, which was dependent upon field strength as well as frequency, occurred at 3- and 20°C, with no effect on Na^+ transport. RF-induced Rb^+ uptake was inhibited by ouabain, which led to the conclusion that the effect was due to a direct electromagnetic field interaction with the plasmalemma active transport enzyme, Na^+/K^+ ATPase. In accordance with the conclusion of Allis and Sinha-Robinson (1987) referred to above, Serpersu and Tsong (1983) indicated that the stimulatory effect on Rb^+ transport was due to an amplitude- and frequency-dependent RF electromagnetic field-induced conformational change in Na^+/K^+ ATPase molecules in red cell membranes. In direct contrast to the results of Allis and Sinha Robinson (1987), who reported inhibition of Na^+/K^+ ATPase activity in membrane fragments exposed to 2.45 GHz RF radiation at field strengths of approximately 90 V/m, Serpersu and Tsong (1983) found stimulation of Na^+/K^+ ATPase activity at a significantly lower frequency (1 kHz) and significantly higher field strength (2 kV/m). The results of Serpersu and Tsong (1983) and Allis and Sinha-Robinson (1987) provide additional evidence of direct frequency-dependent effects of RF radiation on mammalian erythrocytes. Since the apparent target molecule in these studies, Na^+/K^+ ATPase, is ubiquitous in mammalian cells, it may be concluded that RF radiation directly affects all mammalian cells in like manner. Marked qualitative and quantitative differences in cation transport, and other endpoints, detected in cells exposed in vitro to different frequencies and intensities of RF radiation suggest varied responses in in vivo systems exposed under similar conditions, a prediction borne out by extant RF bioeffects data.

Other attempts at determining mechanisms of RF bioeffects have produced negative results. For example, Allis and Sinha (1981) investigated the effect of RF radiation on cell membrane fluidity. They studied the internal viscosity of human erythrocyte membranes exposed to 1 GHz RF radiation at SARs of 0.6-, 2-, and 154 W/kg at temperatures of from 15- to 40°C. There was no detectable RF effect on motional activation energy of membrane fluorescent probes or on lipid phase transitions. Kim et al. (1985) also used membrane fluorescent probes to investigate the effects of 340- or 900 MHz RF radiation on erythrocyte- and erythrocyte-ghost-membranes. Membrane lipid viscosity, the structural state of lipid/protein contact regions, and protein shielding of lipids were decreased during RF exposure. These effects, however, were attributed to indirect thermally-induced changes (Kim et al., 1985). Structural changes in membrane proteins during exposure to RF radiation reported by Ortner, et al. (1981) were also concluded to result from RF-induced heating. Considering these results it may be tentatively concluded that RF-induced alteration in membrane cation fluxes are a consequence of direct interactions with membrane macromolecules, or macromolecular complexes, such as Na^+/K^+ ATPase instead of more generalized alterations in biomembrane lipids or fluidity.

There are somewhat conflicting reports of RF effects on cell surface charge density, as determined by electrophoretic mobility. Ismailov (1977) detected exposure-duration dependent increases or decreases in the electrophoretic mobility of human erythrocytes exposed to 1 GHz RF radiation. A dose-related maximum mobility increase of 23% occurred 30 min after exposure at 37°C at a SAR of 45 W/kg. Human erythrocytes were exposed to 1 GHz RF radiation at 37°C by Bamberger et al. (1981) at SARs of approximately 75- to 150 W/kg. No effect of RF radiation on electrophoretic mobility was detected in this study. Since the SARs in the Bamberger et al. (1987) study were higher than employed by Ismailov (1977), and since Bamberger used a cell concentration 100 times greater than Ismailov (1977), the question remains open regarding RF effects on cell surface charge density.

LEUKOCYTES

Leukocyte exposure to CW and amplitude-modulated RF fields in vitro indicates that specific cell functions are affected. Lyle et al. (1983), for example, detected significant inhibition of allogenic cytotoxicity of target cells (MPC-11) by murine cytotoxic T-lymphocytes when the assay was conducted during exposure to a 450 MHz RF field sinusoidally amplitude-modulated at 60 Hz at an incident intensity of 1.5 mW/cm^2. A similar response was detected when T-lymphocytes were exposed to the RF field prior to assay, suggesting a direct RF field interaction with cytolytic T-lymphocytes. RF-induced cytolytic activity was reversible; T-lymphocytes recovered full activity 12.5 h after exposure. Maximum suppression of T-lymphocyte cytotoxicity occurred as a result of exposure to 450 MHz fields modulated at 60 Hz, relative to 40-, 16-, or 3 Hz modulation. Unmodulated (CW) 450 MHz carrier waves had no effect on cytotoxicity. The authors concluded that the recognition phase of the cytolytic effect was affected by amplitude modulated RF radiation. Mechanisms suggested by Lyle et al. (1983) for effects on T-cell cytotoxicity include field interaction with glycoprotein target cell receptor molecules and/or modulation of critical calcium ion fluxes.

Membrane related modulation of human lymphocytes by RF radiation has been reported at other modulation frequencies. Byus, et al. (1984) exposed human tonsil lymphocytes for periods of up to 60 min. to 450 MHz fields with a peak intensity of 1 mW/cm^2, sinusoidally amplitude modulated at frequencies between 3 and 100 Hz. Total non-cAMP-dependent lymphocyte protein kinase activity decreased to less than 50% of control levels following 15 to 30 min exposures at 16 Hz modulation frequency, whereas there was a smaller decrease following exposure at 60 Hz. Forty-five to 60 min after exposure to continuous RF radiation, protein kinase activity returned to normal values, indicating a "time window" response. No exposure related effects on cAMP-dependent protein kinase activity were detected, suggesting that the effect was due to a transient induction of a persistent reversible molecular state in specific components of the lymphocyte transductive system (Byus et al., 1984).

The effect of RF and microwave radiation on another lymphocyte function, capping of plasma membrane antigen-antibody complexes, was investigated by Sultan et al. (1983a,b). B-lymphocytes were exposed for 30 min to 2.45 GHz microwave radiation at a SAR of 45 W/kg at 37, 41 and 42.5°C, or to 147 MHz RF radiation amplitude modulated by a 9-, 16-, or 60 Hz sine wave at a maximum SAR of approximately 2 W/kg and at temperatures of 37 or 42°C. For unirradiated lymphocytes there was a significant temperature dependent decrease in capping. No difference was detected between exposed or sham exposed cells at any temperature when capping was investigated after RF exposure. There was no significant difference in capping between controls and B-lymphocytes exposed to 2.45 GHz microwaves during capping, when the sample temperatures were both maintained at 38.5°C (Sultan et al., 1983a).

Lloyd et al. (1986) investigated the effects of 20 min exposure of human blood *in vitro* to 2.45 GHz microwaves at SARs of 4 to 200 W/kg. During exposure under conditions of passive temperature control the blood temperature increased from 37 to 40°C. There was no increase in chromosomal damage in RF-exposed blood lymphocytes. Cleary et al. (1987a) investigated the effects of isothermal (37 $\pm$ 0.2°C) RF and microwave exposure on mitogenesis in human peripheral lymphocytes *in vitro*. At SARs to 50 W/kg, a 2 h exposure to 2.45 GHz CW microwave radiation increased PHA-stimulated mitogenesis 3 days after exposure. Simultaneous exposure of blood aliquots to 27 MHz CW RF radiation at the same temperature (37 $\pm$ 0.2°C), at SARs of the same magnitude, caused similar increases in PHA stimulated mitogenesis. Exposure at 50 W/kg or greater, on the other hand, resulted in statistically significant suppression of mitogenesis 3 days postexposure. These effects indicate direct field-induced cellular alterations, not related to RF heating.

Ottenbreit et al. (1981) reported that 2.45 GHz CW microwave radiation at SARs of 250 W/kg or greater inhibited the proliferative capacity of human neutrophil precursor colony-forming cells (CFU-C) following exposures for 15 min at temperatures of 7.0, 22.0, 37.0 or 41.0°C. The irreversible effect was dose-dependent and unrelated to temperature or to the state of the cell cycle. It was noted that CFU-C requires the addition of exogenous glycoprotein stimulators for cell growth. Since the glycoproteins are known to interact with neutrophil membrane receptors, microwave exposure apparently alters membrane receptors, making them less responsive to stimulatory factors. A similar mechanism of interaction could explain the RF-induced suppression of PHA-stimulated mitogenesis reported by Cleary et al. (1987).

Cleary et al. (1985b) investigated the effect of 100 MHz RF radiation on phagocytic activity of rabbit neutrophils (polymorphonuclear leukocytes (PMN)). Neutrophils were exposed *in vitro* for 30 min to CW or amplitude modulated (20 Hz) 100 MHz RF radiation at SARs of 120 to 341 W/kg (induced electric field strengths 250 to 410 V/m) under isothermal (37 $\pm$ 0.2°C) conditions. There was no detectable effect on neutrophil viability or phagocytosis compared to sham-exposed control cells.

NEURAL CELLS

There have been persistent indications that mammalian neurons are one of, if not the most sensitive cell type for RF-induced alterations *in vivo*. This has been the rationale for studies of effects on brain tissue and neural cells *in vitro*. The sensitivity of brain cells to RF radiation *in vitro* is among the highest reported.

Wachtel et al. (1975) and Seaman and Wachtel (1978) reported effects of 1.5- and 2.45 GHz CW and pulse modulated microwaves on the firing rate of neurons from the marine gastropod *Aplysia in vitro*. Unexpected alterations were detected in cells exposed to microwave radiation of SARs between 1 and 100 W/kg. Microwave irradiation reversed the normal temperature dependence of firing rate in 13% of the *Aplysia* pacemaker cells. The rate decreased or fell to zero in response to microwave-induced temperature elevations. The rapid response component, which occurred within 1 s after the start of microwave exposure, was always associated with a decrease in firing rate.

Increased temperature, on the other hand, increased the neural firing rate. At a SAR of approximately 1 W/kg, about 0.1% of the absorbed microwave energy produced a polarizing current across the membrane which resulted in altered firing rates. It was concluded that this was a direct effect of microwave radiation on a neural cell in vitro, an effect reported by other authors.

Hyperpolarization of the resting potential of neurons of the mollusk Helix pomatia as a result of a 1 h exposure to 2.45 GHz CW microwaves at a SAR of approximately 15 W/kg was reported by Arber (1976). By treating the neurons with ouabain after exposure, it was demonstrated that the alteration in hyperpolarization was due in part to inhibition of Na^+/K^+ ATPase, a finding of interest in light of the findings of Allis and Sinha-Robinson (1987) of a similar direct microwave effect on this cation transport enzyme in the erythrocyte membrane in vitro, discussed above. Yamaura and Chichibu (1967) reported effects of 11 GHz microwave radiation at a SAR of 100 W/kg on crayfish and prawn ganglia firing rates. There was a rapid decrease in firing rate during exposure. After exposure the rate rebounded to greater than normal before returning to pre-exposure levels. Elevated temperature cause an increase in firing rate as reported by Wachtel et al. (1975) and Seaman and Wachtel (1978), indicating the effect was not due to RF-induced heating.

Pickard and Barsoum (1981) exposed single giant algal cells from Chara braunii and Nitella flexilis to 250 ms pulse-modulated 0.1 to 10 MHz RF. Fast and slow changes in membrane resting potential were induced by RF exposure. Whereas the slow component was attributed to RF-induced temperature rise, the amplitude of the fast component, which was inversely related to RF frequency, vanished at 10 MHz. Pickard and Barsoum (1981) suggested that the fast component was due to rectification of the RF field by cell membranes, an effect theoretically predicted on the basis of membrane dielectric properties (Franceschetti and Pinto, 1984). Barsoum and Pickard (1982) used similar experimental methods to extend their studies to RF frequencies of 20-300 MHz. They failed to detect membrane rectification at higher frequencies, in agreement with dielectric theory applied to cell membranes. Liu et al. (1982), in addition to studying RF effects on Chara membrane resting potentials, investigated the effects of CW, pulse-modulated and sinusoidally amplitude modulated S-band microwave radiation on amplitude of the action potential, rise and decay time of the action potential, conduction velocity, and cell excitability. Cells maintained under isothermal conditions at 22 ± 0.1°C during exposure showed no consistent or statistically significant microwave-dependent alteration, in basic agreement with the results of Pickard and Barsoum (1981) and Barsoum and Pickard (1982). Dielectric theory predicted no membrane rectification at 2.45 GHz. The results of studies of algal cells thus revealed no direct effects at frequencies of greater than 10 MHz, in apparent disagreement with the results of other studies. This difference may, however, be related to differences in the structure and/or function of plant versus animal excitable cells (Liu et al., 1982).

The effect of RF radiation on rat brain energy metabolism was investigated by Sanders et al. (1984). These studies were conducted with brain tissue per se, rather than cell preparations. A dose-dependent increase in nicotinamide adenine dinucleotide (NADH) fluorescence was detected following a 2 min exposure to 200 or 591 MHz CW radiation, with a threshold electric field strength of 3 to 5 V/m. Exposure of rat brain tissue to the same intensity of 2.45 GHz CW microwave radiation had no detectable effect on energy metabolism. Exposure to 200 or 591 MHz RF radiation decreased brain adenosine triphosphate (ATP); 2.45 GHz exposure had no effect. The energy metabolite creatine phosphate (CP) was suppressed by exposure to 591 MHz radiation, but not by 200 or 2450 MHz, RF radiation of the same SARs. It was proposed that RF radiation induced a direct frequency-dependent effect on specific mitochondrial enzymes and/or on electron transport proteins involved in maintenance of brain cell ATP pools. RF-induced dipole oscillations involving the divalent metal ion at the active site during either catalytic or transport activity was proposed as the mechanism for RF effects on brain energy metabolism (Sanders et al. 1984). This suggested mechanism, involving direct RF interaction with cell biomolecules, is qualitatively similar to that advanced by Allis and Sinha-Robinson (1987) to explain the RF effect on Na^+/K^+ ATPase activity in the erythrocyte membrane, as referred to previously in this section. Additional experimental evidence supporting their hypothesis was provided by Sanders et al. (1985) as a result of studies of the effects of CW, sinusoidal amplitude modulated, and pulsed square-wave modulated 591 MHz radiation on brain energy metabolism.

Statistically significant decreases in incorporation of ^{3}H-thymidine and ^{3}H-uridine in human glioma cells (LN71) in vitro were detected by Cleary et al. (1987b), 3- or 5-days after a 2 h exposure at 37 $\pm$ 0.2°C to 27 MHz CW RF radiation at electric field strength of 160 V/m or greater. Simultaneous exposure of glioma under identical conditions to 2.45 GHz microwave radiation, but at lower electric field strengths (i.e., < 1.3 V/cm), increased ^{3}H-thymidine uptake 3-d postexposure. These results are indicative of a persistent field strength dependent effect of isothermal RF exposure on cellular DNA and RNA synthesis (Cleary et al. 1987b).

RF EFFECTS ON OTHER CELL TYPES

The effects of RF radiation on other cell types and on various cellular endpoints have provided additional evidence of direct cellular responses. Peters et al. (1979) reported effects on the cell cycle in vitro, including cell-type specific inhibition of proliferation following exposure to 2.45 GHz RF radiation. RF radiation altered the growth of synchronized L60T cells as a result of exposure during the miotic (M) and intermiotic (G1) phases of the cell cycle. Exposure of a synchronous L60T cell cultures at four successive 4 h intervals, the time at which the cultures entered the M and G1 phases of the miotic cycle, produced a significant suppression of proliferation.

In vitro methods were used by Balcer-Kubiczek and Harrison (1985) to study the carcinogenicity of RF radiation. C3H/10T½ mouse embryo fibroblasts were exposed in vitro to 4.4 W/kg pulsed 2.45 GHz microwave radiation for 24 h. Latent transformation damage resulted, as indicated by the action of the tumor promoter 1,2-9-tetradecanoyl-phorbol-1,3-acetate (TPA). The potential of microwave radiation to promote cell transformation in vitro is of interest in view of the reports by Szmigielski et al. (1980), Szmigielski et al. (1982) and Kunz et al. (1985), of tumor promotion by microwaves in vivo. Cleary et al. (1989) exposed mouse spermatozoa in vitro for 1h to 27- or 2450 MHz CW RF radiation under isothermal conditions (37 $\pm$ 0.2°C). Exposure to either RF frequency at 50 W/kg or greater caused statistically significant reduction in the ability of irradiated sperm to fertilize mouse ova in vitro. There were no readily detectable exposure effects on sperm morphology, ultrastructure, or capacitation. Reduced fertilization was attributed to a direct effect on spermatozoa rather than to RF-induced heating.

Additional evidence of the sensitivity of mammalian brain cells to modulated RF and lower frequency electromagnetic fields was provided by investigations of effects on Ca^{++} binding to chick brains *in vitro*. It was revealed that low intensity amplitude modulated electromagnetic fields change Ca^{++} binding under specific exposure conditions that involved intensity or modulation frequency biphasic or windowed responses. The windows were found to depend upon magnetic fields having intensities equivalent to geomagnetic fields.

Bawin and Adey (1976), for example, reported a reduction in the release of $^{45}Ca^{2+}$ from chick and cat cerebral tissue exposed *in vitro* to 1, 6, 16, 32, or 75 Hz sinusoidal electric fields. Applied electric field strength thresholds of 10 and 56 V/m were detected for chick and cat brain tissue, respectively. Smaller field induced release also occurred at 5 and 100 V/m. The maximum reduction in $^{45}Ca^{2+}$ release, which was on the order of 12 to 15%, occurred at 6 and 16 Hz.

Frequency and amplitude windows for field-induced release of $^{45}Ca^{2+}$ from chick cerebral tissue also resulted from exposure to VHF and UHF fields amplitude modulated at ELF frequencies. Exposure of chick brain tissue to 8 W/m^2, sinusoidally amplitude modulated 147 MHz RF radiation, caused a statistically significant increase in $^{45}Ca^{2+}$ efflux at modulation frequencies of from 6 to 16 Hz. There was a decreased efflux in response to fields modulated from 20 to 35 Hz (Bawin et al., 1975). Similar effects of sinusoidally modulated 147 MHz RF fields on chick brain tissue were reported by Blackman et al. (1979), who detected a modulation frequency windowed response between 9 and 16 Hz. Effects of these modulated RF fields on calcium efflux were significant only at incident power densities of approximately 10 W/m^2. A similar power density window was detected for chick brain tissue exposed to 450 MHz RF radiation sinusoidally modulated at 16 Hz (Bawin et al., 1978). Increased $^{45}Ca^{2+}$ efflux occurred only at 1.0 and 10 W/m^2, but not at 0.5 or 50 W/m^2. Electric field strengths induced in brain tissue that induced calcium released were on the order of 0.1 V/cm. The ELF alternating magnetic field component was involved in $^{45}Ca^{+2}$ release from chick brain tissue and the effectiveness of the field at a given frequency depended on the local static geomagnetic field intensity (Blackman et al., 1985). Liboff (1985) suggested the interaction of alternating and static magnetic fields with charged ions such as Ca^{+2} may be due to cyclotron resonances. Evidence supporting cyclotron resonance as a mechanism for the effects of ELF electric or magnetic fields on living systems was provided by behavioral studies with rats (Thomas et al. 1986) and diatoms (Smith et al., 1987).

Mechanisms for the ELF field-induced release of Ca^{+2} from brain, or other tissues, remain uncertain. The field-induced efflux of calcium from brain tissue appears to occur from membrane surfaces, rather than intracellular calcium pools (Lin-Liu and Adey, 1982). Although physiological consequences of ELF biphasic or windowed phenomena have not been ascertained, it is obvious that calcium ions are involved significantly in membrane transductive coupling in various neurological, immunological, and endocrinological processes in mammals. Consequently, if windowed responses are induced *in vivo* by exposure to low intensity ELF fields the potential exists for physiological perturbations of undetermined significance. Existing RF safety standards do not take this phenomena into account in formulating safe levels of exposure to modulated fields.

OTHER FIELD EFFECTS

Evidence of nonthermal effects of low frequency electric and magnetic fields, as well as high amplitude pulsed electric fields, has also been reported. The results of such studies, although largely unexplained, lead to the general conclusion that biological systems are more sensitive to perturbation by electromagnetic fields than previously recognized.

Effects of DC electric fields on cell membrane receptors were reported by several investigators. Orida and Poo (1978) detected electrophoretic movement and localization of acetylcholine receptors in *Xeropus* embryonic muscle cell membranes exposed to 1 kV/m DC electric fields for 1.5 h. Receptors accumulated at the cathodal side of the cells and formed stable metabolically independent receptor aggregates. The migration of concanavalin A receptors on the cell surface of cultured xenopus myoblasts in response to pulsed DC electric fields was reported by Lin-Liu and Adey (1984). Receptor migration occurred within 150 min in an electric field having time-averaged field strength of 30 V/m. Asymmetric receptor distributions were detected within 5 min after application of a 150 V/m pulsed electric field. The degree of receptor asymmetry was a function of the duration of field exposure, pulse width, frequency, and electric field strength. Following attainment of steady-state receptor distributions, pulsed fields having the same time-averaged field strength produced effects equivalent to DC fields, indicating that the primary mechanism was electrophoresis.

The orientation and growth rate of embryonic neurites in explanted nervous tissue and neurons in vitro is affected by DC electric fields (Jaffe and Poo, 1979). In spatially uniform pulsed fields or focal DC or pulsed fields, neurites preferentially oriented forward the cathodic pole (Patel and Poo, 1982). Time-averaged pulsed fields induced affects similar to equivalent intensity DC fields. The mechanisms responsible for field-induced neurite orientation have not been determined. Electrophoretic redistribution of growth-controlling membrane surface glycoproteins has been advanced as a possible explanation for this effect (Patel and Poo, 1982).

The site of the majority of the cellular effects of electromagnetic fields is purportedly the plasma membrane. Genomic effects have, however, been reported as well. Goodman and Henderson (1986) exposed salivary gland cells of *Sciara coprophilia* to quasi-rectangular, asymmetric pulsed magnetic fields, or sinusoidally amplitude-modulated magnetic fields. Regardless of the magnetic field waveform, transcriptional activity was increased. Field enhanced transcription occurred principally at chromosome loci that were normally active during the developmental stage at which the cells were exposed, indicating magnetic field-induced enhancement rather than alteration of transcriptional patterns. Induced transcriptional activity was maximum in response to exposure to a 72 Hz, 1.15 millitesla (mT) magnetic induction.

Cleary et al. (1988) used a chicken tendon explant model system to investigate the effects of ELF, low amplitude, unipolar, square wave pulsed electric fields on fibroplasia in vitro. A 1 Hz electric field with a pulse duration of 1 ms and a time-averaged current density of 7 mA/m^2 induced maximum (32%) increase in fibroblast proliferation in explants exposed for 4 days. Exposure to the same field at a current density of 1.8 mA/m^2 had no effect on cell proliferation, whereas exposure to current densities greater than 10 mA/m^2 inhibited proliferation and collagen synthesis, without affecting noncollagen protein synthesis. Fibroplasia was increased significantly in explants oriented parallel to applied electric fields that induced current densities of 3.5 or 7 mA/m^2, but there was no effect on explants exposed to these same fields but oriented perpendicular to the field (Cleary et al., 1988).

Exposure of cells *in vitro* to high amplitude microsecond duration pulsed DC electric fields induces a different class of effects. Exposure to such fields can irreversibly alter membrane permeability leading to osmotic lysis. Electric field-induced membrane permeability alterations have been reported by several investigators (Sale and Hamilton, 1968; Reimann, et al., 1975; Kinosita and Tsong, 1977a,b; Teissie et al., 1982; Smith and Cleary, 1983). The electric field strength and pulse duration thresholds for the induction of membrane alterations are dependent on cell type. Experimental evidence, as well as theoretical models, suggest that membrane alterations occur when the transmembrane potential is increased by approximately 1V. Applied electric fields that induce such transmembrane potentials are of such short duration that membrane permeation can occur under conditions not involving detectable Joulian heating; the effect thus is nonthermal in nature. Reimann et al., (1975) proposed dielectric breakdown as the general mechanism for this effect.

CONCLUSIONS

Review of the bioeffects literature, especially the results of *in vitro* cellular studies provides convincing evidence that RF radiation, and other types of electric or magnetic fields, can alter living systems via direct nonthermal mechanisms, as well as via heating. Examples of such direct effects of RF radiation reviewed here include: (1) alteration of membrane Na$^+$/K$^+$ ATPase activity, (2) changes in membrane cation fluxes, (3) decrease in non-cAMP-dependent protein kinase activity, (4) allogenic T-lymphocyte inhibition, (5) biphasic effects on lymphocyte proliferation, (6) effects on neutrophil precursor membrane receptors, (7) altered rates of synthesis of DNA and RNA in glioma cells, (8) changes in brain cell energy metabolism, (9) alteration in the firing rates and resting potentials of neurons, (10) cell cycle alteration, and (11) cell transformation.

Obviously, the results of such *in vitro* studies cannot be directly applied to the determination of safety standards for humans. Considered as a whole, however, these *in vitro* cellular effects are not inconsistent with reported effects of RF radiation on laboratory animals or humans. This introduces the distinct possibility that effects detected in *in vitro* systems may be induced *in vivo* as well, as would be anticipated on a theoretical basis because the same biomolecular interaction mechanisms must be involved in either case. Consequently, the current basis for setting Western safety standards, namely of restricting exposure to minimize thermally-induced alterations, must be reexamined in light of the emerging evidence of nonthermal RF bioeffects. The major impediment to the meaningful utilization of *in vitro* data in such reconsideration remains a lack of understanding of basic mechanisms of interaction and effects. The potential of *in vitro* studies to yield physical insight indicates the strong need for continuation of such studies. Pending the results of such studies, as well as the results of additional *in vivo* studies, current safety standards must be viewed as interim health protection expedients.

REFERENCES

Allis, J.A. and Sinha, B.L. (1981). Fluorescence depolarization studies of red cell membrane fluidity. The effect of exposure to 2.0-GHz microwave radiation. Bioelectromagnetics 2,13-22.

Allis, J.W. and Sinha-Robinson, B.L. (1987). Temperature-specific inhibition of human red cell Na^+/K^+ ATPase by 2,450-MHz microwave radiation. Bioelectromagnetics, 8,203-212.

Arber, S.L. (1976). Effect of microwaves on resting potential of giant neurons of mullusk Helix Pomatia. Electronnaya Obrabotka Materialov, 6,78-79.

Balcer-Kubiczek, E.K. and Harrison, G.H. (1985). Evidence for microwave carcinogenesis in vitro. Carcinogenesis 6,859-864.

Bamberger, S., Keilmann, F., Storch, F., Roth, G. and Ruhenstroth-Bauer, G. (1981). Experimental results contradicting claimed 1009-MHz influence on erythrocyte mobility. Bioelectromagnetics 2,85-88.

Barsoum, Y.H. and Pickard, W.F. (1982). Effects of electromagnetic radiation in the range 20-300MHz on the vacuolar potential of characean cells Bioelectromagnetics 3,193-201.

Bawin, S.M. and Adey, S. (1976). Sensitivity of calcium binding in cerebral tissue to weak environmental electrical fields oscillating at low frequency. Proc. Natl. Acad. Sci. U.S.A. 73, 1999-2003.

Bawin, S.M., Adey, W.R., Sabbot, I.M. (1978): Ionic factors in release of $^{45}Ca^{2+}$ from chicken cerebral tissue by electromagnetic fields. Proc. Natl. Acad. Sci. USA 75:6314-6318.

Blackman, C.F., Elder, J.A., Weil, C.M., Benane, S.G., Eichinger, D.C., and House, D.E. (1979). Induction of calcium-ion efflux from brain tissue by radiofrequency radiation: Effects of modulation frequency and field strength. Radio Sci. 14(6S), 93-98.

Blackman, C.F., Benane, S.G., Rabinowitz, J.R., House, D.E. and Joines, W.T., (1985). A role for the magnetic field in the radiation-induced efflux of calcium ions from brain tissue in vitro. Bioelectromagnetics 6,327-337.

Byus, C.V., Lundak, R.L., Fletcher, R.M. and Adey, W.R. (1984). Alterations in protein kinase activity following exposure of cultured human lymphocytes to modulated microwave fields. Bioelectromagnetics. 5,341-351.

Claret, M., Garay, R. and Giraud, F. (1978). The effect of membrane cholesterol on the sodium pump in red blood cells. J.Physiol. (London). 274,247-263.

Cleary, S.F. (1973). Uncertainties in the evaluation of the biological effects of microwave and radiofrequency radiation. Health Physics. 25:387-395.

Cleary, S.F., Liu, L.M., Nickless, F. and Smith, G. (1979). Effects of pulsed DC fields on mammalian blood cells. In "The Mechanisms of microwave Biological Effects". (Taylor, L.S. and Cheung, A.W., eds.) Electrical Engineering Dept. and Inst. for Physical Science and Technology, Division of Mathematical and Physical Sciences and Engineering and the Dept. of Radiation Therapy, School of Medicine, Univ. of Maryland, College Park, MD, p.16

Cleary, S.F., Garber, F. and Liu, L.M. (1982). Effects of X-band microwave exposure on rabbit erythrocytes. Bioelectromagnetics 3,453-466.

Cleary, S.F. (1983). Microwave radiation effects on humans. Bioscience. 33,269-273.

Cleary S.F., Liu, L.M., Garber, F. (1985a): Erythrocyte hemolysis by radiofrequency fields. Bioelectromagnetics. 6:313-322.

Cleary, S.F., Liu , Garber F. (1985b): Viability and phagocytosis of neutrophils exposed in vitro to 100 MHz radiofrequency radiation. Bioelectromagnetics 6:53-60.

Cleary, S.F., Liu, L.M. and Merchant, R. (1987a) Comparison of the effects of isothermal microwave and RF radiation of human lymphocyte mitogenesis in vitro. Bioelectromagnetics Society Ninth Annual Meeting Abstracts. June 21-25, 1987, Portland, OR

Cleary, S.F. Liu, L.M. and Merchant, R. (1987b). RF and microwave effects on proliferation, DNA and RNA synthesis in glioma cells in vitro. Bioelectromagnetics Society Ninth Annual Meeting Abstracts, June 21-15, 1987, Portland, OR

Cleary, S.F. (1988). Biological Effects of Nonionizing Radiation. In Encyclopedia of Medical Devices and Technology, Vol. 1, (E. Webster, ed.) John Wiley & Sons, New York, N.Y.

Cleary, S.F., Liu, L.M. Diegelmann, R.F. (1988) Modulation of tendon fibroplasia by exogenous electric currents. Bioelectromagnetics 9:183-194

Cleary, S.F., Liu, L.M., Graham, R., East, J. (1989): In vitro fertilization by mouse spermatozoa exposed isothermally to radiofrequency radiation. Boelectromagnetics (to be published in 1989).

D'Andrea, J.A., Emmerson, R.Y., Bailey, C.M., Olsen, R.G. and Gandhi, O.P. (1985). Microwave radiation absorption in the rat: Frequency dependent SAR distribution in body and tail. Bioelectromagnetics. 6(2),199-206.

D'Andrea, J.A., Emmerson, R.Y., DeWitt, J.R.D. and Gandhi, O.P. (1987). Absorption of microwave radiation by the anesthetized rat: Electromagnetic and thermal hotspots in body and tail. Bioelectromagnetics 8, 385-396.

Franceschetti, G. and Pinto, I. (1984). Cell membrane nonlinear response to applied electromagnetic fields. IEEE Trans. Microwave Theory Tech. MIT-32, 166-169.

Fisher, P.D., Poznarsky, M.J. and Voss, W.A.G. (1982). Effect of microwave radiation (2450MHz) on the active and passive components of $^{24}Na^+$ efflux from human erythrocytes. Radiation Res. 92,411-422.

Gandhi, O.P. (1975). Conditions of strongest electromagnetic power deposition in man and animals. IEEE Trans. Microwave Theory and Techniques. MTT-23,1021-1029.

Goodman, R. and Henderson, A.S. (1986). Sine waves enhance cellular transcription. Bioelectromagnetics. 7,23-29.

Ismailov, E.Sh. (1977). Effect of ultrahigh frequency electromagnetic radiation on the electrophoretic mobility of erythrocytes. Biofizika 22,443-498, transl. in Biophysics. 22,510-516 (1978).

Jaffe, L., Poo, N. (1979): Neurites grow faster towards the cathode than the anode in a steady field. J. Exp. Zool. 209:115-128.

Kim, Y.A., Fomenko, B.s., Agafonova, T.A., and Akoev, I.G. (1985). Effects of microwave radiation (340 and 900MHz) on different structural levels of erythrocyte membranes. Bioelectromagnetics. 6, 305-312.

Kinosita, K., Tsong, T.Y. (1977a): Hemolysis of human erythrocytes by a transient electric field. Proc. Natl. Acad. Sci. U.S.A. 74:1923-1927.

Kinosita, K., Tsong, T.Y. (1977b): Formation and resealing of pores of controlled sizes in human erythrocyte membranes. Nature. 268:438-441.

Kunz, L.L., Johnson, R.b., Thompson, D., Crowley, J., Chou, C.K. and Guy, A.W. (1985). Effects of Long-Term Low-Level Radiofrequency Radiation Exposure on Rats. USAF Sch. Aerosp. Med. Rep. SAM-TR-85-11, Vol. 8, Brooks Air Force Base, Texas.

Liboff, A.R. (1985). Geomagnetic cyclotron resonance in living cells, J.Biol. Phys. 13,99-102.

Liburdy, R.P. and Penn, A. (1984). Microwave bioeffects in the erythrocyte are temperature and pO_2 dependent: Cation permeability and protein shedding occur at the membrane phase transition. Bioelectromagnetics. 5,283-291.

Liburdy R.P. and Vanek, P.F. (1985). Microwaves and the cell membrane II. Temperature, plasma, and oxygen mediate microwave-induced membrane permeability in the erythrocyte. Radiation Res. 102,190-205.

Lin-Lui, S. and Adey, W.R. (1982). Low frequency amplitude modulated microwave fields change calcium efflux rates from synaptosomes. Bioelectromagnetics 3, 309-322.

Lin-Liu, S., Adey, W.R. (1984): Migration of cell surface concanavalin A receptors in pulsed electric fields. Biophys. J. 45:1211-1217.

Liu, L.M., Nickless, F.G., Cleary, S.F. (1979). Effects of microwave radiation on erythrocyte membranes. Radio Science. 14,109-115.

Liu, L.M., Cleary S.F. (1988): Effects of 2.45GHz microwaves and 100MHz radiofrequency radiation on liposome permeability at the phase transition temperature. Bioelectromagnetics 9:249-258.

Liu, L.M., Garber, F. and Cleary, S.F. (1982). Investigation of the effects of continuous-wave, pulse-and amplitude-modulated microwaves on single excitable cells of Chara Corallina. Bioelectromagnetics. 3,203-212.

Lyle, D.B., Schechter, P., Adey, W.R., and Lundale, R.L. (1983). Suppression of T-lymphocyte cytotoxicity following exposure to sinusoidally amplitude-modulated fields. Bioelectromagnetics. 4,281-292.

Lloyd, D.C., Saunders, R.D. Moquet, J.E. and Kowalczuk, C.I. (1946). Absence of chromosomal damage in human lymphocytes exposed to microwave radiation with hyperthermia. Bioelectromagnetics 7, 235-237.

Olcerst, R.B., Belman, S., Eisenbud, M., Mumford, W.W. and Rabinowitz, J.R. (1980). The increased passive efflux of sodium and rubidium from rabbit erythrocytes by microwave radiation. Radiation Res. 82, 244, 256.

Orida, N., Poo, M. (1978): Electrophoretic movement and localization of acetylcholine receptors in the embryonic muscle cell membrane. Nature. 275:31-35.

Ortner, M.J., Galvin, M.J., McRee, D.I., Chignell, C.R. (1981): A novel method for the study of fluorescence probes in biological material during exposure to microwave radiation. J. Biochem. Biophys. Methods. 5: 157-167.

Ottenbreit, M.J., Lin, J.C., Inoue, S. and Peterson, W.D. (1981). In vitro microwave effects on human neutrophil precursor cells (CFU-C). Bioelectromagnetics. 2,203-215.

Patel, N.B., Poo, M. (1982): Orientation of neurite growth by exracellular electric fields. J. Neurosci. 24:483-496.

Peters, W.J., Jackson, R.W. and Iwano, K. (1979). Effects of controlled electromagnetic radiation on the growth of cells in tissue culture. J. Surgical Res. 27,8-13.

Pickard, W.F. and Barsoum, Y.H. (1981). Radiofrequency bioeffects at the membrane level: Separation of thermal and athermal contributions in the Characeae. J. Membrane Biol. 61,39-54.

Reimann, F., Zimmermann, V., Pilvat, G. (1975): Release and uptake of hemoglobin and ions in red blood cells induced by dielectric breakdown. Biochim. Biophys. Acta. 394:449-462.

Roelofsen, B. (1981). The (non)specificity in the lipid-requirement of calcium and sodium plus potassium-transporting adenosine triphosphatases. Life Sci. 29, 2235-2247.

Sale, A.J.H, Hamilton, W.A. (1968): Effects of high electric fields on micro-organisms. III Lysis of erythrocytes and protoplasts. Biochim. Biophys. Acta. 16:37-43.

Sanders, A.P. Joines, W.T. and Allis, J.W. (1984). The differential effects of 200, 591, and 2450MHz radiation on rat brain energy metabolism. Bioelectromagnetics. 5,419-433.

Sanders, A.P., Joines, W.T., Allis, J.W. (1985). Effects of continuous-wave, pulsed, and sinusoidal-amplitude-modulated microwaves on brain energy metabolism. Bioelectromagnetics. 6,89-97.

Seaman, R.L. and Wachtel, H. (1978). Slow and rapid responses to CW and pulsed microwave radiation by individual Aplysia pacemakers. J.Microwave Power, 13,77-86.

Serpersu, E.H. and Tsong, T.Y. (1983). Stimulation of a ouabain-sensitive Rb^+ uptake in human erythrocytes with an external electric field. J.Membrane Biol. 74,191-201.

Smith, G.K., Cleary, S.F. (1983): Effects of pulsed electric fields on mouse spleen lymphocytes in vitro. Biochim. Biophys. Acta. 763:325-331.

Smith, S.D., McLeod, B.R., Liboff, A.R. and Cooksey, K. (1987). Calcium cyclotron resonance and diatom mobility. Bioelectromagnetics 8, 215-227.

Sultan, M.F., Cain, C.A. and Tompkins, W.,A.F. (1983a). Effects of microwaves and hyperthermia on capping of antigen-antibody complexes on the surface of normal mouse B lymphocytes. Bioelectromagnetics. 4,115-122.

Sultan, M.F., Cain, C.A. and Tompkins, W.A.F. (1983b). Immunological effects of amplitude modulated radiofrequency radiation: B-lymphocyte capping. Bioelectromagnetics. 4,157-165.

Szmigielski, S., Szydzrinski, A., Peitraszek,A. and Bielec, M. (1980). Acceleration of cancer development in mice by long-term exposition to 2450-MHz microwave fields. In "URSI International Symposium Proceedings Ondes Electromagnetiques et Biologie", (A.J. Berteaud and B.Servantie, eds.). Paris, France. pp. 165-169.

Szmigielski, S., Szymdzinski, A. Pietraszek, A., Bielec, M., Janiak, M. and Wrenbel, J.K. (1982). Accelerated development of spontaneous and benzopyrene-induced skin cancer in mice exposed to 2450-MHz microwave radiation. Bioelectromagnetics. 3,179-191.

Teissie, J., Knutson, V.P., Tsong, T.Y., Lane M.S. (1982): Electric pulse-induced fusion of 3T3 cells in monolayer cultures. Science 216:537-538.

Thomas, J.R., Schrot, J., Liboff, a.R. (1986). Low-intensity magnetic fields alter operant behavior in rats. Bioelectromagnetics 7,349-357.

Wachtel, H., Seaman, R. and Joines, W. (1975). Effects of low-intensity microwaves on isolated neurons. Ann.N.Y.Acad. Sci. 247,46-62.

Yamaura, I. and Chichibu, S. (1967) Super-high frequency electric field and crustacean ganglionic discharges. Tohoku J. Exp. Med., 93, 249-259.

ELECTROMAGNETIC FIELDS AND NEOPLASMS

Stanislaw Szmigielski[1] and Jerzy Gil[2]

[1] Department of Biological Effects of Nonionizing Radiations, Center for Radiobiology and Radiation Safety, 128 Szaserow, 00-909 Warsaw, Poland
[2] MMA Postgraduate Medical School, 128 Szaserow 00-909 Warsaw, Poland

INTRODUCTION

Cancer morbidity has been growing rapidly in the first few decades and it is believed that to a large extent the increase in incidence of neoplasms is related to growing exposure to harmful environmental factors that may either induce (carcinogens) or promote neoplastic transformation and/or growth and spread of tumors. Recognition of environmental factors related to cancer morbidity and investigation of their carcinogenic potency and mechanisms of interaction with respect to neoplasms is amongst the most important prophylactic tasks leading to lowering of cancer morbidity and protection of the population.

Electromagnetic fields (EMFs) are a relatively new but a rapidly intensifying ecologic factor. For millions of years the biological life on the earth has been developing and undergoing evolution under influence of complex natural electric, magnetic and electromagnetic fields generated by the solar radiation and electric phenomena in the atmosphere. The exposure was relatively weak, as the natural intensities of EMFs at the range 1 KHz-300 GHz does not exceed the order of 10^{-8}-10^{-9} W/m^2. The situation has changed dramatically for the last 50 years with introduction of various devices generating radiofrequencies (RFs) and microwaves (MWs), first for communications and navigation and later for multiple industrial and household purposes. Steadily increasing occupational groups and subpopulations living in certain areas (e.g., close to power lines and stations, TV/radio broadcasting antennas, military bases, etc.) are being exposed to EMF intensities that exceed the natural fields of the earth by a few orders of magnitude. Measurements of RF/MW fields in certain areas of United States accessible to the public revealed intensities of 10^{-2}-10 W/m^2. For instance, the area south of Syracuse, New York, where numerous transmitters are located, showed fields of 10^{-2} W/m^2 for a space of several square miles in the vicinity of antennas (Cohen, 1978). Near Portland, Oregon and Seattle, Washington the readings of 1 to 7 W/m^2 were recorded (Marino, 1988), while in Denver, Colorado in an area accessible to the public, values of up to 10 W/m^2 were observed near the RF/MW antennas, though the indoor levels were somewhat lower and ranged from 0.5 to 6 W/m^2 (Marino, 1988).

At present only the second and third generation of human beings are exposed to man-made EMFs and long-term effects of these exposures are still difficult to foresee. Despite numerous experimental investigations and epidemiological studies (for review, see Elder and Cahill, 1984) it is still not possible to prove the existence and character of any specific molecular, cellular or systemic damage that may be related to long-term exposure to weak EMFs. However, a variety of behavioral, neurological, endocrine, hemato-immunological or reproductive abnormalities have been reported under these conditions. Reports of these abnormalities, including the phenomenon regarded as adaptive or compensatory reactions, have led some authors to recognize EMFs as a nonspecific biological stressor that is detected by the nervous system and may be a risk factor for a variety of diseases, depending on susceptibility or reactivity to stress and/or individual genetic predisposals of the exposed host (Marino, 1988).

Surprisingly, very little attention has been paid to the possible links between exposure to EMFs and development of neoplastic diseases. Despite various causal and anecdotal reports on appearance of various neoplasms in relatively young men servicing and/or repairing radar devices (McLaughin, 1953; Zaret, 1977; Dwyer and Leeper, 1978) and one experimental study indicating development of unspecified leukosis in mice exposed for about one year to 9270 MHz microwaves (Prausnitz and Süsskind, 1962) the cancer-related aspects of EMFs were until the late seventies never a subject of deeper investigations. Although there were no evidences and arguments to support this view, a prevalent opinion of the bioelectromagnetics experts was that EMFs were not carcinogenic. In 1979 Weltheimer and Leeper published their first epidemiological report indicating increased mortality from neoplasms in children living close to high-voltage power lines and related this phenomenon to exposure to magnetic fields. At the same time Szmigielski et al. (1980) presented (during the URSI Symposium on Bioeffects of Electromagnetic Waves in Paris) preliminary results on accelerated development of spontaneous (genetically determined breast cancer) and chemically induced (benzopyrene-evoked skin tumors) neoplasms in mice exposed chronically (for a few months) to nonthermal fields of 2450 MHz. The above two reports gained much interest and doubts both from the bioelectromagnetics community and from the public and encouraged some groups of scientists to undertake epidemiologic or experimental studies on the relation of EMFs to cancer development (Tables I through III).

The aim of the present review is to summarize the available data on increased risk of neoplasms in subjects exposed to EMFs, indicate the unresolved issues that will guide future investigations, discuss the possible mechanisms of cancer-promoting potency of EMFs as nonspecific stressors and finally, to answer the question whether or not at the present state of knowledge, the postulated promotional effect of public and/or occupational exposure to EMFs should influence the safety criteria.

ROLE OF ENVIRONMENTAL AND OCCUPATIONAL FACTORS IN CARCINOGENESIS

Carcinogenesis has come recently to be viewed as a multistage process involving both the cellular effects, leading to neoplastic transformation followed by uncontrolled growth and, on the other hand, the host's antineoplastic systems (e.g., immune surveillance mechanisms, nonspecific cytotoxicity, interferon production determining the individual status of nonspecific antineoplastic resistance) that effectively protects healthy organisms against development of neoplastic disease. Individually, genetically programmed susceptibility to transforming agents, frequent exposure to carcinogens (e.g., tobacco smoking) or transient lowering of the antineoplastic resistance unbalances the physiological homeostasis and may result in development of neoplastic disease.

For simplicity, the process of cancer development can be divided into three major stages: initiation, occurring at the cell nucleus as a result of different endogenous or exogenous factors leading to appearance of oncogens and appearance of single transformed cells; promotion, being in fact a form of privilege for survival and/or growth of transformed cells by factors acting directly on cellular metabolism or membrane function (e.g., phorbol esters) or by factors influencing the host's antineoplastic resistance (e.g., immunosuppressive agents, stressors); and finally, cocarcinogenesis, defined as mechanisms facilitating formation of neoplastic tumors, including their vascularization and spread.

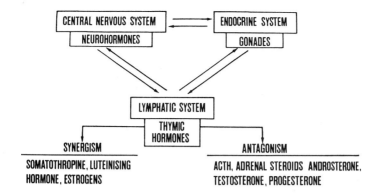

Fig. 1. Physiologic systems regulating and controlling homeostasis of the organism and reaction to external stimuli.

Initiation of carcinogenesis, appearance of cells with genetic mutations and/or chromosomal abnormalities as well as neoplastic transformation of single cells occurs quite frequently in human beings and the incidence of these events may be increased with age or due to habits, (diet, alcohol and tobacco consumption), action of stressors and ability to cope with stress (Sklar and Anisman, 1981) or exposure to environmental or occupational factors. Integration of homeostasis of the human organism is provided by the neuro-endocrine-lymphatic network (Fig. 1) that via numerous control and regulatory mechanisms ensures proper function of all systems of the organism as well as triggering of adaptive and compensatory reactions to a variety of situations occurring in everyday life as well as for chronic exposure to environmental and occupational toxins and stressors. There are numerous experimental observations that neurohormones can influence immune reactions and vice versa, and thymic hormones, being regulators of immunity, act synergistically or antagonistically with the endocrine and nervous system activities (for review, see Stein et al., 1982). Function of the neuro-endocrine-lymphatic network and efficiency of the three systems are genetically determined and thus, it is logical to assume individual differences in susceptibility to endogenous and exogenous factors influencing the above network (Fig. 1), including the ability to cope with acute and chronic stressors. With the present state of knowledge we are not able to measure and categorize the efficiency of the whole network in one parameter, but it was demonstrated that the distribution of immune status (anti-infective resistance) in human subjects considered as healthy follows the normal Gaussian curve with presence in the population of genetically determined cases of lowered and elevated resistance. The problem of different individual susceptibility to environmental or occupational factors was only rarely discussed, although it may be important for assessment of cancer risk. In case of EMFs a problem of individuals being hypersensitive to weak electric and magnetic fields has been reported only recently (Choy et al., 1987) and needs further elaboration.

In general, environmental factors, including EMFs, may influence carcinogenesis or development of neoplastic diseases in a variety of ways (Fig. 2).

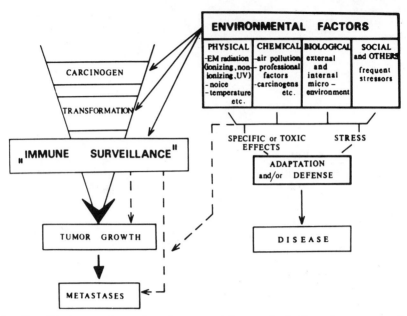

Fig. 2. Possible interactions of environmental factors, including electromagnetic fields with development of neoplastic disease and the immune surveillance mechanisms.

Environmental factors evoke both specific effects on various levels of biological organization (molecular, cellular, systemic) and the concomitant nonspecific stress reaction. For defined factors the two components of biological reactions are expressed at certain levels (e.g., for EMFs the stress reaction predominates). Specific effects and the stress reaction trigger adaptation and/or defense (compensatory) mechanisms (Fig. 2), being a form of the host's defense again harmful effects caused by the factor. When capacity of the adaptation and compensatory mechanisms is exceeded, the factor-related disease (environmental or occupational disease) occurs (Fig. 2). In case of EMFs the specific disease does not appear; nevertheless, both the specific effects (i.e., increased neoplastic transformation rate, alteration of certain enzymes related to carcinogenesis, see Table III) and the triggered adaptation/compensatory mechanisms (i.e., the stress-related stimulation of hypophysial-adrenal axis, lowering of antineoplastic resistance) may influence the process of carcinogenesis and development of clinically-diagnosable neoplastic disease. Precise mechanisms of interaction of EMFs with the above processes cannot be defined at this time, but the experimental investigations at the cellular level in vitro (see Table III) and immunosuppression caused by long-term exposure to MW fields (for recent review, see Szmigielski, et al., 1988) suggest tumor promotion and not direct carcinogenic potency of EMFs.

CANCER-RELATED ASPECTS OF EXPOSURE TO ELECTROMAGNETIC FIELDS

Potential carcinogenicity has been periodically discussed in relation to RF/MW exposure from 1953, when McLaughin (1953) listed various forms of leukemias as one of the possible effects of occupational exposure to radar radiation. This was followed by anecdotal and scientifically or epidemiologically unproven reports on appearance of a variety of forms of neoplasms in relatively young (30-40 years of age) men servicing or repairing radars for several years and exposed to strong MW fields (Zaret, 1977; Dwyer and Leeper, 1978). In Poland we have also observed an unexpected concentration of cases of neoplastic diseases in units where microwave devices were constructed, tested or repaired. Among personnel of one of the military schools of telecommunications, four cases of hemato-immunologic neoplasms (two lymphomas, one chronic myelocytic leukemia and one acute myelocytic leukemia) were diagnosed in a short period of time, while in one department of a technical university, where microwave devices were built and tested experimentally, two cases of pancreatic cancer occurred at the same time in men aged 38 and 52 years, respectively (unpublished information).

The above observations raised many doubts among both American and Polish bioelectromagnetics experts and authorities responsible for safety standards and were not accepted as related to occupational exposure to MWs. In 1982 Lester and Moore reported significantly higher incidence of cancer mortality in United States counties with Air Force bases, compared to counties without an Air Force base and related the observed differences to chronic MW exposure of the populations living near the bases. Again, the results and conclusions have not been accepted widely (Polson and Merritt, 1985) but the original authors still support their views (Lester and Moore, 1985).

During the years 1980 through 1988 over 30 reports on epidemiological data concerning elevated risk of cancer in human beings exposed to environmental or occupational EMFs, including both RF/MW radiation and power frequency electric and magnetic fields as well as several experimental investigation on cancer-related aspects have been published. As the above data cause much stir, anxiety and doubts both among the bioelectromagnetics community and the general public, it appears to be the proper time to evaluate critically the validity of the available materials, stress the uncertainties concerning the relation of cancer to EMF exposures and suggest guidelines for further investigation. The available data pertain to occupational and environmental (public) exposure to human beings, and experimental observations on cells and animals irradiated under controlled conditions of exposure.

Occupational Exposure

Most of the epidemiologic data indicating increased cancer morbidity (or mortality) related to occupational exposure to EMFs are based on retrospective analyses of death certificates of cancer victims or subpopulations living in an investigated area or employed in one factory or type of industry (Table I). Some of these studies are uncontrolled (no comparable control group or population), while other (Coleman et al., 1983; Flodin et al., 1986; Szmigielski et al., 1988) are case-controlled. In nearly all reports the certificates were analyzed according to profession (or occupation type) listed in the certificate and type of neoplasms diagnosed for the victim. This appears to be the weakest point of all the available reports (Table I), as the listed profession or occupation type may allow only for conjecture concerning exposure to EMFs, its frequency and modulation characteristics, intensity and duration, which introduces uncontrolled bias in the population. It should be stressed, however, that the above uncertainties concerning real exposure to EMFs among individuals of a certain occupational group may "blur" a real relation of cancer development to exposure to EMFs, as subjects with uncertain, short-lasting and/or weak exposure are analyzed together with those being exposed much more intensively. Analyses of groups by profession, however, will allow conclusions on relationship of cancer incidence to type of EM radiation (power frequencies, radiofrequencies, microwaves) that might be an important factor for future preventive and safety considerations.

Despite all of the above reservations, the data listed in Table I seem convincing that certain occupational groups working with electric and electronic devices are subjected to increased risk of hematoimmunologic and neurologic neoplasms, although the relation of this increase to exposure to EMFs may be only postulated at the present time. Amongst the groups at the increased risk are the electricians, electronic and electrical engineers and technicians, linemen, power station workers, computer, radio and telephone operators, etc. Thus, a postulated exposure to electric and magnetic fields of power frequency with focus of some authors on strong static and low frequency magnetic fields (Milham, 1982; Stern, 1986) arouses suspicion as to the cause of increased risk of leukemias. However, in two studies (Milham, 1985; Szmigielski et al., 1988) a relationship to exposure to RFs and MWs is also postulated.

Table I. Increased risk of neoplasms related to occupational exposure to electromagnetic fields.

Type of neoplasms	Occupation, exposure type, basic results	Source of information Population or group analyzed	Reference
Leukemia	Radar workers, causal observations	Military personnel	McLaughin, 1953.
Various neoplasms	Causal and anecdotal cases, 2 brain tumors among 18 radar servicemen, 5 cancers among 17 radarmen, 3 cancers among 8 repairmen of radars	Personnel working with radar, service and repair	Zaret, 1977
Leukemia	1. Aluminum industry workers exposed to strong magnetic fields (20 cases v. 10.6 expected). 2. Electricians, linemen, power station workers, electronic and electrical engineers (136 cases, 92 expected, p<0.05)	Analysis of 438,000 death certificates in Washington state during 1950-1979.	Milham, 1982
Leukemia, acute forms	Electronic and electric technicians and engineers, 23 observed, 13.3 expected, p < 0.05	Death certificates of white males in Los Angeles	Wright, et al., 1982
Leukemia	10 categories of occupations related to electronic and/or electric industry 96.5 expected, p<0.05	Death certificates in southeastern England tries, 113 observed,	Coleman, et al. 1983
Leukemia	Electricians (odd ratio 3.0) and welders (odd ratio 3.83) employed in the shipyard	53 cases of neoplasms among white males in Portsmouth Naval Shipyard	Stern, 1986
Acute myeloid leukemia	Electrical technicians, welders, computer and telephone technicians (odd ratio 3.8)	Case-control study of death certificates	Flodin, et al., 1986
Leukemia, various types	1. Electronic and electric technicians (odd ratio 2.3. 2. Electrical workers 3. Electrical engineers, radio and telephone operators	537 leukemias in England and Wales (1973) New Zealand Wisconsin 1963-1978	McDowell, 1982 Pearce, et al. 1985 Calle, et al.1985

Table I. (Continued)

Type of neoplasms	Occupation, exposure type, basic results	Source of information, population or group analyzed	Reference
Leukemia	Amateur short-wave radio operators, 24 leukemias observed, 12.6 expected, $p<0.01$	1691 death certificates of male members of American Radio Relay League	Milham, 1985
Brain and other neurological neoplasms	1. Electricians, electrical engineers, linemen (50 cases of glioma and astrocytoma observed, 18 expected, $p<0.01$) 2. Doubled incidence of brain and eye neoplasms	951 cases of brain tumors among white male residents of Maryland, 1969-1982; Reanalysis of data of Robinette et al., 1980	Lin, et al. 1985
Brain tumors	Electronic and electrical industry workers	Death certificates in England	Coggan, et al. 1986
Brain tumors and leukemia	Increased incidence in powerline workers but normal levels in power station operators	Cancer registry data in Sweden	Tornqvist, et al. 1986
Neoplasms of eyeball	Electronic and electrical industry workers	Death certificates in England and Wales (1962-1977)	Swerdlow, 1983
Various neoplasms	Excess of cancer morbidity by 15% in electronic industry workers	Swedish cancer environmental registry	Vagero and Olin, 1983
Colorectal cancer	A three-fold increase of intestinal neoplasms in high voltage powerline workers	Cancer registry data in Canada	Howe and Lindsay, 1983
Leukemia, lymphoma, stomach, colorectal and skin neoplasms	A three-fold increase of general cancer morbidity with seven-fold increase of certain hemato-immunologic neoplasms	Case-control study of cancer morbidity in Polish military career personnel exposed and nonexposed to RFs and MWs (1971-1980)	Szmigielski, et al., 1988

All the data available at the present time have appeared from retrospective epidemiologic analyses and thus are valid only for the populations being investigated and for the time periods covered by the analyses. Any single study does not allow for generalization or conclusions about causitive relationship, but if the same information comes from numerous sources and locations, as is the case of people working in electric/electronic industry (Table I), it allows one to accept the phenomenon, in spite of the lack of firm proof as to its cause. Thus, it is the present authors' feeling that the available data should not influence the safety standards and levels of EMFs but that there is an urgent need for prospective epidemiological studies on relationship of exposure to EMFs (type of radiation, intensity of exposure, period of employment) and on influence of other occupational and environmental (including life style and habits) factors combined with EMFs on risk of cancer.

Upon scrutinizing the data listed in Table I it appears that two types of neoplasms predominate -- leukemias and brain tumors, both relatively rare and both insufficiently specified in terms of the present diagnostic criteria. Only in one study (Howe and Lindsay, 1983) based on cancer registry in Canada, a three-fold increase of the intestinal neoplasms in high voltage powerline workers was noted. Surprisingly, lung cancer, which is a frequent type of neoplasm in male population in Europe and United States and strongly dependent on environmental factors and life habits, did not show increased frequency in workers of the electric/electronic industries. Similar trends were observed in the retrospective epidemiological study of cancer morbidity in the whole population of Polish career military personnel aged 20-59 years (divided into age groups by decades) subdivided into subjects exposed occupationally to RF/MW radiation and non-exposed controls (Szmigielski et al., 1988). All cases of neoplasms registered during the decade of 1971-1980 were analyzed according to the age of victims, exposure to RF/MWs (including period of exposure) and organ localization. Cancer morbidity in subjects exposed to RF/MWs (all age groups) was triple that of the non-exposed group (192.2 versus 64.2 per 100,000 subjects per year, respectively (Fig. 3). Statistical differences were found for incidence of neoplasms of blood-forming and lymphatic systems ($p<0.01$) and for stomach, colorectal and skin neoplasms, including melanoma ($p<0.05$). The large difference in incidence of hemopoietic and lymphatic neoplasms (all together listed probably as "Leukemias" in studies presented in Table I) suggested differentiation and analysis of single forms of these neoplasms (Fig. 4). In the whole population the incidence of hemopoietic and lymphatic neoplasms was relatively low (about 4 per 100,000 at the age of 20 - 39 years, about 12 per 100,000 for 40 - 49 and about 33 per 100,000 for 50 - 59 years) and the increase in the RF/MW-exposed subjects regarded only to lymphosarcomas, lymphomas, chronic and acute myelocytic leukemias, but not to malignant lymphogranulomatosis, chronic lymphocytic leukemia and plasmocytoma.

Despite the statistical significance of differences in cancer morbidity in personnel exposed and non-exposed to EMFs and high correlation of the morbidity with period of exposure, we do not relate these findings directly to interaction of the radiation with human organism at any level (Szmigielski et al., 1988) until the prospective studies of the same population planned for 1986-1990 are completed and more information on exposure intensities is obtained.

Interestingly, for those who follow the lead of the Russian scientists working on biological effects of EMFs and their advocacy of very low safety levels, no reports on increased incidence of neoplasms related to public or occupational exposure to EMFs have appeared from U.S.S.R. and no investigations in this field were undertaken until now (Akoev, personal information during the Comecon Countries Symposium on Biological Effects and Safety Rules of Nonionizing Radiations in Puschino, U.S.S.R., October, 1987).

Environmental (public) Exposure

Epidemiological data concerning the increased risk of neoplasms related to environmental exposure to EMFs are much less convincing and documented than the earlier discussed indices in personnel exposed occupationally. The two basic reports (Wertheimer and Leeper, 1979, 1982 and Lester and Moore, 1982) that postulated a relation

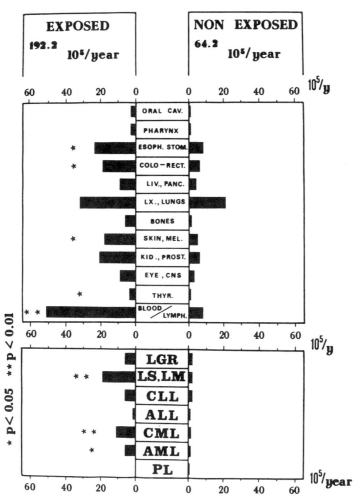

Fig. 3. Incidence of neoplasms (per 100,000 subjects yearly) in personnel exposed and non-exposed (controls) to radio-frequency and microwave radiation. Top histograms -- organ localization of malignancies; Oral cavity -- neoplasms of oral cavity, Esoph. Stom. -- esophageal and stomach; Liv., panc. -- liver and pancreatic; lx -- laryngeal; mel. -- melanoma; kid., prost. -- kidney and prostatic, CNS -- central nervous system; thyr. -- thyroid; lymph. -- lymphatic organs. Bottom histograms -- morbidity rate for specified types of malignancies originating from blood-forming and lymphatic organs; LGR -- malignant lymphogranulomatosis (Hodgkin's disease), LS -- lymphosarcoma; LM -- lymphoma (non-Hodgkin); CLL -- chronic lymphocytic leukemia; ALL -- acute lymphoblastic leukemia; CML -- chronic myelocytic leukemia; AML -- acute myeloblastic leukemia; PL -- plasmocytoma (plasma cell leukemia). (Reprinted from Szmigielski et al., 1988).

AGE GROUP	Population		LGR	Ly Sa Lymp.	CLL	ALL	CML	AML	PL
	TOTAL	EXPOSED / NON EXPOSED							
20-29	3.6	26.3 / 2.7	18.8 / 2.1	0.3		8.8 / 0.3	8.8		
30-39	3.8	29.7 / 3.0	0.9	0.3	9.9 / 0.3		19.8 / 1.2	0.3	
40-49	11.8	81.3 / 9.9	11.6 / 2.5	46.5 / 4.2	11.6 / 1.4		11.6 / 0.3	11.6 / 1.8	
50-59	32.7	117.6 / 29.6	3.0	58.8 / 8.9	8.9		29.4 / 1.1	29.4 / 5.9	
TOTAL	8.8	50.8 / 7.4	6.0 / 1.8	18.3 / 2.2	6.1 / 1.3	3.0 / 0.1	12.2 / 0.5	6.1 / 1.1	2.2
χ^2		70.01	2.88	29.85	1.11	14.65	45.32	7.52	0.06
P		<0.01	NS	<0.01	NS	<0.01	<0.01	<0.05	NS
RISK FACTOR		6.7		8.3		7.8	9.6	5.5	

Fig. 4. Morbidity rate of hemopoietic and lymphatic malignancies (number of cases per 100,000 subjects per year) in personnel exposed occupationally to microwave and radio-frequency radiations and in unexposed controls in different age groups (20-29, 30-39, 40-49 and 50-59 years). For explanation of abbreviations, see Fig. 3, bottom histograms. Note the largest differences in the 40-49 year age group and earlier appearance of malignancies in the RF/MW-exposed group. (Reprinted from Szmigielski et al., 1988).

between long-term exposure to power frequency fields or to microwaves at areas near high voltage lines or radar antennas and higher incidence of neoplasms (mainly leukemias) in residents of these areas, were both criticized (Fulton et al., 1980; Polson and Merrit, 1985) and contradicted by another report postulating no increase of leukemia rate in children living near power lines (Fulton et al., 1980) (Table II). Measurements of intensities of electric and magnetic fields of power frequency in schools and homes located near power lines revealed values well below those suggested even in the most stringent safety standards (Daley et al., 1985). In Oregon, United States, power frequency electric fields did not exceed 3 - 5 V/m inside schools and 1- 170 V/m outside schools, while magnetic fields ranged about 1 mGauss inside and 35 - 75 mGauss outside the schools (Daley et al., 1985). In the school with higher intensities of electric and magnetic fields four cases of neoplasms were diagnosed in a fourth-grade class within a period of the school year. The authors (Daley et al., 1985) concluded that the low intensities of EMFs could not be responsible for the four cases of different neoplasms but did not offer another explanation. In fact, it is difficult to accept the relationship of extremely weak EMFs for relatively short period of exposure (few years) with increased risk of neoplasms, unless the subjects are individually hypersensitive to EMFs. In none of the reports (Table II) the problem of hypersensitivity of cancer victims among residents near power lines or evaluation of their immune status prior to development of neoplasms is discussed and it may be assumed that this problem was not analyzed.

Table II. Increased risk of neoplasms related to environmental (public) exposure to electromagnetic fields.

Type of neoplasms	Basic results of analysis, data on exposure and relation to cancer morbidity	Source of information, population or group analyzed	Reference
Various neoplasms	Doubled rate of cancer morbidity in a rural area at Syracuse, New York, U.S.A., traversed by numerous high voltage power lines and containing 20 RF/MW antennas	Uncontrolled study of cancer registry (by residency)	Becker, 1977
Various neoplasms	Increased mortality from neoplasms in children living near power lines; the phenomenon related to magnetic fields of power lines	Controlled study of death certificates in Denver area (1950-1973)	Wertheimer and Leeper, 1979
Leukemia, lymphoma and neurological neoplasms	Doubled the expected incidence of hematologic and neurologic neoplasms in children (101 cases observed, 55 in control group) and in adults (438 v. 372, respectively) living near power lines	Controlled study of death certificates in Denver area	Wertheimer and Leeper, 1982
Leukemia (no increase in risk)	Normal incidence of leukemia in children living near power lines	Controlled study of death certificates in Rhode Island	Fulton, et al.,
Various neoplasms	Higher incidence of cancer mortality in population living in U.S. counties with Air Force bases, compared to mortality in counties without Air Force bases; differences related to long-lasting exposure to microwaves (radar radiation). Critical reevaluation of the results, uncertainty concerning relation to radars	Controlled study of death certificates in U.S. counties Reevaluation of the above data	Lester and Moore, 1982; Lester and Moore, 1985 (reply to Polson and Merritt, 1985). Polson and Merritt, 1985

Wertheimer and Leeper (1982) reported 101 cases of various hemato/lymphatic and neurologic malignancies in children living near power lines, as against the statistical expectation of 55 for this group (ratio about 2.0). Referring to data listed in Table I on cancer incidence in personnel exposed occupationally to EMFs, where exposures are normally more intense and longer-lasting, ratios between number of observed and expected cases range 0.15 - 3.8 and often ratios of about 2.0 are seen, being similar to that found in children exposed environmentally.

Very similar ratios of observed vs. expected cases of cancer under radically different conditions of exposure are difficult to explain. This dose-effect dependency raises further doubts on relationship of neoplasms to environmental exposure to EMFs.

EXPERIMENTAL INVESTIGATIONS

Experimental data supporting the correlation of long-term exposure to EMFs with process of neoplastic transformation of cells in vitro and in vivo and with growth of neoplastic tumors in animals are still scarce and fragmentary (Table III). Most of the reports listed in Table III seem to be a byproduct of experiments planned for more general noncarcinogenic aims. For example, Prausnitz and Süsskind (1962) observed a variety of reactions in mice exposed during 59 weeks to short daily sessions of 9270 MHz MW hyperthermia (100 mW/cm^2, 4.5 minutes daily) and found occurrence of "leukosis" in part of the irradiated animals. These authors did not characterize the phenomenon in detail and did not continue their observations. On the basis of numerous experiments with modulated RFs and MWs Adey and his group (1981) have more recently concluded that pulse and amplitude modulation of the carrier wave at the 1-100 Hz frequency appears to be a prime determinant of the nature of interaction at the cellular level. When this group continued their experiments and switched to immunocompetent and neoplastic cell cultures irradiated in vitro in modulated MW fields, they obtained results important for both immunologic and cancer-related aspects of interaction of EMFs at the cellular level -- alterations of activity of two enzymes that are essential for nucleic acid and protein turnover as well as for neoplastic transformation of cells -- histone protein kinase and ornithine decarboxylase (Byus et al., 1984, 1986). These alterations, although transient, occurring only at certain modulations of the carrier wave (phenomenon of "windowing") and disappearing quickly after completion of the irradiation of cells, are no doubt evoked by the radiation and indicate that weak EMFs may under certain conditions alter cell metabolism and membrane function in a way that may be relevant to neoplastic transformation of the cells.

In 1985 Balcer-Kubiczek and Harrison reported data indicating strong synergism between phorbol esters (known promoters of carcinogenesis, active in vitro and in vivo) and nonthermal MW fields in increasing the transformation rate of cells cultured in vitro and treated formerly with benzopyrene or x-rays as initiating agents. Briefly, the experiments performed by these authors indicate that exposure of murine embryonic fibroblasts (established cell line in vitro) to nonthermal MW fields (2450 MHz, pulse modulated at 120 pulses/sec, pulse duration 83 μ sec, SAR 4.4 mW/g) yields a statistically significant (1.6 to 3.5-fold) increase in neoplastic transformation rate compared to transformation of cells irradiated with x-rays and treated with TPA (phorbol ester). However, exposure to MWs had no effect on transformation induced by carcinogen (benzopyrene or x-rays) in the absence of promoter (TPA). The authors (Balcer-Kubiczek and Harrison, 1985) conclude that MWs induce latent transformation damage which may be revealed by action of promoters of carcinogenesis (e.g., TPA). In general, the results indicate that cell membrane may be a sensitive target for the influence of low-level EMFs.

The above-discussed experiments in vitro strongly indicate the possibility of interaction of weak EMFs with membrane function and neoplastic transformation of cells in vitro. Thus, it is surprising that this line of research has developed slowly, in spite of the availability of valid techniques for studying consecutive steps of neoplastic transformation in vitro, including potency of promoters. Virtually nothing is known on dependence of the above-discussed effects on frequency of EMFs, intensity of the applied fields and action of electric and magnetic fields of power frequency.

Experimental investigations with tumor-bearing animals irradiated with EMFs under controlled conditions of exposure (Table III) appear also to be at the preliminary stage and developing slowly. In all of the listed experiments only the final (integrated) effect in form of appearance and/or growth of neoplastic tumors was controlled without any closer insight into single stages of carcinogenicity or function of the immune surveillance mechanisms. Thus, it is not surprising that discussion of this problem (see also Szmigielski et al., 1988) is in most cases referred to papers from our group (Szmigielski et al., 1980, 1982; Szudzinski et al., 1982) and to the reports from Guy's investigations (Guy et al., 1985; Kunz et al., 1985), both needing confirmation and continuation.

Table III. Experimental investigations of influence of electromagnetic fields on carcinogenesis.

Observed Effects	Experimental Material	Conditions of Exposure	Reference
Inhibition of histone protein kinase activity	Culture human lymphocytes	450 MHz, 1.0-1.5 mW/cm^2, sinusoid-	Byus, et al., 1984
Increase of ornithine decarboxylase activity; Potentiation of stimulation caused by promoter (phorbol ester)	Cultured hepatoma cells	450 MHz 1 mW/cm^2 sinusoidally modulated at 3-100 Hz	Byus, et al. 1986
Increase of transformation rate of cells treated with carcinogen and promoter	Cultured murine, embryonic fibroblasts in vitro	2450 MHz pulsed (120 pulses/sec. pulse width 83 μsec.), SAR 4.4 mW/g	Balcer-Kubiczek and Harrison, 1985
Increased frequency of leukosis	Mice	9270 MHz, 100 mW/cm$_2$ (thermogenic field), 4.5 min. daily 59 weeks	Prausnitz and Süsskind,
Delayed development of transplantable sarcoma in young mice irradiated prenatally (in utero)	Pregnant mice and offspring	2450 MHz, 20 min. daily, SAR 35 mW/g (thermogenic field), days 11-14 of gestation, implantation of sarcoma in 16-day-old offspring	Preskorn et al., 1978
Accelerated development and growth of benzopyrene-induced skin cancer	Mice	2450 MHz, 5-15 mW/cm^2, SAR 3-4 mW/g, 2-hr. daily during 1-3 months	Szmigielski et al., 1980, 1982; Szudzinski et al., 1982
Increased number of spontaneous neoplasms (no predominance for type)	Rats	2450 MHz, pulsed (800 pulses/sec., pulse 10 μsec.) 0.48 mW/cm^2 throughout lifetime	Guy et al., 1985; Kunz et al., 1985
Unchanged growth rate of transplantable melanoma	Mice	2450 MHz, 2-hr. daily, 10 mW/cm^2, after implantation of melanoma	Isatini et al., 1988

Guy and his coworkers (Guy et al., 1985; Kunz et al., 1985) reported increased number of spontaneous malignant tumors appearing in rats exposed for their life-span under specimen-free conditions to weak MW fields (pulsed 2450 MHz, 800 pulses/sec., pulse duration 10 μsec., power density 0.48 mW/cm^2, SAR of 0.15 - 0.4 mW/g, 23 hours daily). A group of 200 animals was used for irradiation, another 200 rats served as sham-exposed controls. In the exposed group 54 malignancies were found, while in the sham-exposed controls the corresponding number was only 23, the difference being statistically significant at $p < 0.05$. However, no preference for a specific type of cancer was found; in particular number of hemato-immunologic malignancies did not differ between the exposed and nonexposed groups.

Findings from our group (Szmigielski et al., 1980, 1982; Szudzinski et al., 1982) have been recently reevaluated and discussed (Szmigielski et al., 1988). Briefly, daily (two-hour) exposures of mice to continuous 2450 MHz MWs at field power densities 5-15mW/cm^2 (SAR 2-3 and 6-8 mW/g) continued for three to six months resulted in accelerated appearance and growth of skin neoplasms induced by benzopyrene, suggesting a tumor-promoting activity of MWs. Interestingly, chronic stress from confinement of mice (unexposed) used as a positive control, gave a similar acceleration of growth of benzopyrene-induced skin tumors as was observed in mice exposed to MW fields at 5 mW/cm^2. An additional finding was lowering of natural antineoplastic resistance of mice (measured by number of neoplastic lung colonies after i.v. injection of viable sarcoma cells) after exposure to 2450 MHz MWs at 5 - 15 mW/cm^2 (2 hours daily) for one to three months. Again, a similar lowering of antineoplastic resistance has been observed in mice exposed to chronic stress from confinement. In further experiments Szmigielski et al. (1988) have confirmed the acceleration of chemically-induced tumors by chronic exposure to MWs for two other carcinogens -- di-ethyl-nitrosoamine (DENA) and methylcholantrene.

A relationship between stress and cancer have jberen widely investigated in experimental animals (for review, see Sklar and Anisman, 1981; Bammer, 1981), but general conclusions are still difficult to draw. Both enhancement and inhibition of tumor induction and growth have been observed, depending on type of stressors, duration of stress situation, ability to cope with stress and housing conditions of animals. Chronic physical stress has tended to be associated with inhibition of tumor growth. However, chronic psycho-social stressors are rather responsible for enhancement of development of neoplasms. Thus, it is not possible to relate our findings on acceleration of chemically-induced tumors by chronic exposure to MWs directly to the chronic stress situation, even though the chronic stress from confinement, applied as a control situation in our studies, gave a similar acceleration of tumor growth.

Recently a growth of transplantable melanoma in mice exposed to MWs has been studied (Isatini et al., 1988). The authors implanted B 16 melanoma cells in mice and after implantation started daily exposures of the animals to nonthermal 2450 MHz MW fields. The exposures lasted for the whole period of tumor growth. No differences in tumor growth, final mass of tumors and survival have been found between the exposed and control group of animals. This study indicates that neoplastic cells already transformed and passively implanted to new host are not influenced by nonthermal MW fields and further growth of transplantable tumors is not affected by exposure to MWs. The situation here is different from chemical induction of neoplasms or spontaneous tumors, where first induction and promotion of carcinogenesis occur and transformation of cells is followed by growth and spread of neoplasms, partially controlled by the immune system. The findings of Isatini et al., (1988) compared to the earlier discussed acceleration of chemically-induced and spontaneous neoplasms in animals exposed to MWs suggest that chronic exposure to EMFs influences the early stages of carcinogensis (initiation, promotion) and not growth and spread of transformed cells, but this needs further investigations. It has to be stressed also that transplantable tumors in mice grow relatively fast and lead to death of animals in a few weeks. Thus, a period of exposure to MWs, in this case limited to two to four weeks, is significantly shorter than that used in investigations on development of chemically-induced or spontaneous neoplasms.

In summary, results of experimental investigations available at present provide only limited data, though documented on different models, that exposure to low-level EMFs influence under certain conditions the complicated process of carcinogenesis with a postulated direct (at the cellular level) or indirect (resulting from suppression of immunity) tumor-promoting activity. This field therefore remains open for good research projects with adequate exposure data, free from possible artifacts.

SUMMARY AND CONCLUSIONS

Recent reports on the possibility of increased risk of neoplastic disease, particularly of "leukemias" related to long-term occupational exposure to EMFs have caused concern and have led to misinterpretations and unfounded generalizations both among bioelectromagnetic scientists and the general public. Review of available epidemiological and experimental results reveals that generally incoherent and insufficient data have been used to draw conclusions regarding the cause and effect relationships. On the other hand, all facts indicate that something distressing and not yet understandable occurs in this field.

Numerous epidemiological data arising from a variety of sources and geographical locations indicate that certain professions and types of occupation connected with exposure to EMFs (personnel of the electric and electronic industries, workers of high voltage power lines and power stations, radar and radio operators and repairmen) are subject to increased risk of neoplasms, although a cause of this risk and its severity still remain unclear. The present reports of epidemiological data indicating the increased risk of neoplasms in certain professions allows one to accept existence of the phenomenon and there is no single report denying this. However, relationship to EMF exposure and other occupational factors, combination of EMFs and psychosocial stresses or life habits and other possible bases remains an open question and needs further investigations. For example, age of cancer victims among electric/electronic workers compared to controls or general population was analyzed only superficially. Incidence of neoplasms varies markedly with age and if the subjects live longer, no doubt they will have a higher rate of neoplasms. On the other hand, tumor-promoting factors not being carcinogenic, per se, may accelerate development of spontaneous neoplasms that would be manifested by early appearance of clinically diagnosable cancers. It has to be stressed also that while discussing increased rates of neoplasms one operates with very small figures (a few to several cases per 100,000). Thus, two problems should be considered:

-- Increased risk may refer evenly to all members of the population, but the risk may be assessed as "tolerable" in terms of population and costs of advancing civilization.

-- In the population there are individuals that are exceptionally sensitive to the applied factors and these individuals develop neoplasms with enormously high odds and increase the rate for whole population. Recognition and elimination of sensitive individuals would lower the population rate to normal values.

At present there are no data allowing us to account for the above two problems, but both possibilities should be kept in mind for future epidemiologic investigations.

Data suggesting relationship of neoplasms to environmental (public) exposure to power frequency fields or to microwaves are much less convincing than the discussed observations for subjects with occupational exposure. Not all epidemiological analyses support the phenomenon of increased risk of neoplasms related to public exposure and, furthermore, the field intensities measured for the geographical locations where the observed populations live are much below levels considered as biologically significant. The only logical explanation, if we assume reality of the postulated link of neoplasms to public exposure to EMFs, seems to be the presence of exceptionally sensitive individuals (e.g., subjects with genetically-lowered status of natural resistance).

Experimental investigations have been performed only on few selected schedules of exposure, mostly to MWs pulsed or modulated in a variety of manners and the results cannot be generalized for all EMFs and are not directly relevant to safety standards. Nevertheless, it is documented for a variety of exposure conditions that weak MW fields may influence cell metabolism leading to alterations in nucleic acid and protein synthesis pathways in a manner relevant to neoplastic transformation of cells. Whether or not this is a general phenomenon related to EMF exposure and, still more, whether or not these alterations occur also in vivo, remains an open question and needs further investigations. In animals exposed to MW fields for a long period of time (up to several months) faster development of spontaneous and chemically-induced neoplasms was reported from two laboratories, but again generalization of this phenomenon is not possible until more information is obtained. Surprisingly, experiments on carcinogenicity and growth of neoplasms have not been performed for exposure to electric or magnetic fields of power frequency and this gap needs to be filled.

In general, the available epidemiological and experimental data on relationship of neoplasms to chronic exposure to EMFs are still fragmentary but indicate importance of the problem and urgent need for further investigations. There is no doubt that groups of personnel with occupational exposure to EMFs display increased morbidity of neoplasms and prospective epidemiological investigations of these groups should be undertaken with individual analysis of all newly-occurring cases of neoplasms and a search for potential individuals being exceptionally sensitive to EMFs. Efforts should be made also to establish as precisely as possible the type, duration and intensity of EMF exposure of all cancer victims in the occupationally exposed groups.

The data available at present are too fragmentary and uncertain for influencing the safety levels and guidelines, both the strict standards operating in Eastern European countries and those less stringent being in force in the United States and in Western European countries. Simply, nobody can at present judge what intensities of EMFs and what frequencies of the radiation may be considered as safe, in terms of cancer risk and how to protect occupational groups and the general public from the possible hazard. The only advice we can offer at present is the general rule to avoid unnecessary exposure to factors that might be potentially harmful.

REFERENCES

Adey, W. R., 1981, Ionic nonequilibrium phenomena in tissues interactions with electromagnetic fields, In "Biological Effects of Nonionizing Radiation," K. H. Illinger, ed., ACS Symposium Series, Washington, 157:271.

Balcer-Kubiczek, E. K. and Harrison, G. H., 1985, Evidence for microwave carcinogenesis in vitro, Carcinogenesis 6:859.

Bammer, K., 1981, Stress, spread and cancer, In "Stress and Cancer," K. Bammer and B. H. Newberry, eds., C. J. Hogrefe, Inc., Toronto.

Becker, R. O., 1977, Microwave radiation, N. Y. State J. Med. 77:2172.

Byus, O. V., Kartun, K., Pieper, S., and Adey, W. R., 1986, Microwaves act as cell membranes alone or in synergy with cancer-promoting phorbol esters to enhance ornithine decarboxylase activity, Bioelectromagnetics, 7:432.

Byus, O. V., Lundak, R. L., Fletcher, R. M., and Adey, W. R., 1984, Alterations in protein kinase activity following exposure of cultured human lymphocytes to modulated microwave fields, Bioelectromagnetics, 5:341.

Calle, E. E., and Savitz, D. A., 1985, Leukemia in occupational groups with presumed exposure to electrical and magnetic fields, New Engl. J. Med., 286:1475.

Choy, R., Monro, J., and Smith, C., 1987, Electrical sensitivity in allergy patients, Clinical Ecology, 4:93.

Coggan, D., Pannett, B., Osmond, C., and Acheson, E. D., 1986, A survey of cancer and occupation in young and middle-aged men. II. Nonrespiratory cancers, Brit. J. Industr. Med., 43:381.

Cohen, J., 1978, "Report to the Town Board of the Town of Onondaga," Onondaga, New York.

Coleman, M., Bell., J., and Skeet, R., 1983, Leukemia incidence in electrical workers, Lancet, i:982.

Daley, M. L., Morton, W. E., Cartier, V., Zajac, H., and Benitez, H., 1985, Community fear of ionionizing radiation: A field investigation, IEEE Trans. Biomed. Eng. BME-32:246.

Dwyer, M. J., and Leeper, D. B., 1978, "A current literature report on the carcinogenic properties of ionizing and nonionizing radiation. II. Microwave and radiofrequency radiation," DHEW Publ., NIOSH Report No. 78-134, Washington.

Elder, J. H., and Cahill, D. F., eds., 1984, "Biological effects of radiofrequency radiation," US EPA Report No. 600/8-83-026 F, EPA Research Triangle Park, North Carolina.

Flodin, U., Fredricksson, M., Axelson, O., Persson, B., and Hardell, L., 1986, Background radiation, electrical work and some other exposures associated with acute myeloid leukemia in a case-referent study, Arch. Environ. Health 41:77.

Fulton, J. P., Cobb, S., Preble, L., Leone, L., and Forman, E., 1980, Electrical wiring configurations and childhood leukemia in Rhode Island, Am. J. Epidemiol. 111:292.

Guy, A. W., Chou, C. K., Kunz, L. L., Crawley, J., and Krupp, J., 1985, Effects of low-level radiofrequency radiation exposure on rats. Vol. 9, Summary, Report No. USAF/SAM-TR-85-64, USAF School of Aerospace Med., Brooks Air Force Base, Texas.

Howe, G. R., and Lindsay, J. P., 1983, A followup study of ten-percent sample of the Canadian labor force. J. Nat. Cancer Inst. 70:37.

Isatini, R., Hosni, M., Deschaux, P., and Pacheco, H., 1988, B 16 melanoma development in black mice exposed to low-level microwave radiation, Bioelectromagnetics 9:105.

Kunz, L. L., Johnson, R. B., Thompson, D., Crowley, J., Chou, C. K. and Guy, A. W., 1985, Effects of long-term low-level radiofrequency radiation exposure in rats. Vol. 8, Evaluation of longevity, cause of death and histopathological findings, Report USAF School of Aerospace Med. (AFSC), Brooks Air Force Base, TX 78 235-5301.

Lester, J. R. and Moore, D., 1982, Cancer mortality and Air Force bases. J. Bioelectricity 1:77.

Lester, J. R. and Moore, D., 1985, Reply to "Cancer mortality and Air Force bases": A reevaluation. J. Bioelectricity 4:121.

Lin, R. S., Dischinger, P. C., Condee, J., and Farrell, K. P., 1985, Occupational exposure to electromagnetic fields and the occurrence of brain tumors. J. Occup. Med. 27:413.

Marino, A. A., 1988, Environmental electromagnetic energy and public health, In "Modern Bioelectricity, A. A. Marino, ed., M. Dekker, New York and Basel.

McDowall, M. E., 1983, Leukemia mortality in electrical workers in England and Wales, Lancet i:246.

McLaughlin, J. R., 1953, "A survey of possible health hazards from exposure to microwave radiation," Hughes Aircraft Corporation, Culver City, California.

Milham, S., Jr., 1982, Mortality from leukemia in workers exposed to electrical and magnetic fields. New Engl. J. Med. 307:249.

Milham, S., Jr., 1985, Silent keys: Leukemia mortality in amateur radio operators. Lancet i:812.

Pearce, N. E., Sheppard, R. A., Howard, J. K., Fraser, J., and Lilley, B. M., 1985, Leukemia in electrical workers in New Zealand. Lancet i:811.

Polson, P., and Merritt, J. H., 1985, Cancer mortality and Air Force bases. A Reevaluation. J. Bioelectricity 4:121.

Prausnitz, S., and Süsskind, C., 1962,Effects of chronic microwave radiation on mice. IRE Trans. Biomed., Electron. 9:104.

Preskorn, S. H., Edwards, D., and Justesen, D. R., 1978, Retarded tumor growth and longer longevity in mice after irradiation by 2450 MHz microwaves. J. Surg. Oncol. 10:483.

Sklar, L. S. and Anisman, H., 1981, Contributions of stress and coping to cancer development and growth, In: "Stress and Cancer," K. Bammer and B. H. Newberry, eds., C. J. Hogrefe, Inc., Toronto.

Stein, M., Keller, S. E., and Schleifer, S. J., 1982, The roles of brain and neuroendocrine system in immune regulation -- potential links to neoplastic diseases, In: "Biological Mediators of Behaviour and Disease: Neoplasia," S. M. Levy, ed., Elsevier Biomedical, New York, Amsterdam, Oxford.

Stern, F. B., 1986, A case-control study of leukemia at a Naval nuclear shipyard, Am. J. Epidemiol. 23:980.

Swerdlow, A. J., 1983, Epidemiology of eye cancer in adults in England and Wales 1962-1977, Am. J. Epidemiol. 18:294.

Szmigielski, S., Szudzinski, A., Pietraszek, A., and Bielec, M., 1980, Acceleration of cancer development in mice by long-term exposition to 2450 MHz microwave fields, In: "Ondes Electromagnetiques et Biologie," A. J. Berteaud and B. Servantie, eds., URSI, Paris.

Szmigielski, S., Szudzinski, A., Pietraszek, A., Janiak, M., and Wrembel, J. K., 1982, Accelerated development of spontaneous and benzopyrene-induced skin cancer in mice exposed to 2450 MHz microwave radiation, Bioelectromagnetics 3:179.

Szmiegielski, S., Bielec, M., Lipski, S., and Sokolska, G., 1988, Immunological and cancer-related aspects of exposure to low-level microwave and radiofrequency fields, In: "Modern Bioelectricity," A. A. Marino, ed., M. Dekker, New York and Basel.

Szudzinski, A., Pietraszek, A., Janiak, M., Wrembel, J., Kalczak, M., and Szmiegielski, S., 1982, Acceleration of the development of benzopyrene-induced skin cancer in mice by microwave radiation, Arch. Dermatol. Res. 274:303.

Tornqvist, S., Norell, S., Ahlbom, A. and Knave, B., 1986, Cancer in the electric power industry, Brit. J. Indust. Med. 43:212.

Vagero, D., and Olin, R., 1983, Incidence of cancer in the electronics industry: Using the new Swedish Cancer Environment Registry as a screening instrument, Brit. J. Indust. Med. 40:188.

Wertheimer, N. and Leeper, E., 1979, Electrical wiring configurations and childhood cancer, Am. J. Epidemiol. 109:273.

Wertheimer, N. and Leeper, E., 1982, Adult cancer related to electrical wires near the home, Int. J. Epidemiol. 11:345.

Wright, W. E., Peters, J., and Mack, T., 1982, Leukemia in workers exposed to electrical and magnetic fields, Lancet ii:1160.

Zaret, M. M., 1977, Potential hazards of Hertzian radiation and tumors, New York State J. Med. 77:146.

WORLDWIDE PUBLIC AND OCCUPATIONAL RADIOFREQUENCY

AND MICROWAVE PROTECTION GUIDES

Martino Grandolfo*, and Kjell Hansson Mild**

* National Institute of Health, Physics Laboratory
and INFN-Sezione Sanita', Rome, Italy
** National Institute of Occupational Health, Umea, Sweden

INTRODUCTION

Efforts to establish a scientific data base for setting radiofrequency (RF) and microwave (MW) standards mark their 36th anniversary this year. In April 1953, a conference was held in Bethesda, Maryland, to assess the state of scientific knowledge on RF/MW bioeffects. At the time, the relevant scientific literature numbered less than 100 items, and the scientists assembled at the meeting called for more research, setting in motion the RF/MW bioeffects research effort that has been ongoing, with varying degrees of intensity, ever since.

The two nations that have made by far the greatest contributions to the study of possible effects of exposure to radiofrequency and microwave energies are the USA and the USSR. The well-known wide disparity between the exposure limits of the USA and of the USSR has, therefore, not unreasonably occasioned distrust and sometimes alarm among people who are, or who could be exposed to radiowaves.

In response to this, there has been intense activity, both at the international and national level, on the evaluation of biological effects literature and assessment of health hazards of human exposure to non ionizing electromagnetic radiation. The health authorities of an increasing number of countries become involved in the development of recommendations, regulations or technical advice to limit the exposure of the workers and the general public and they are looking to the international organizations for some guidance in this respect. As a matter of fact, the methodology of protection against ionizing radiation which enabled the safe development of its industrial and medical applications all over the world is often quoted as an example, and every one has in mind the invaluable help provided by the international organizations to resolve the complex scientific and administrative problems encountered in the early management of radiation protection.

A standard is a general term incorporating both regulations and guidelines and can be defined as a set of specifications or rules to promote the safety of an individual or group of people (Repacholi, 1983). Absolute assurances are rarely if ever practicable and specifying maximum permissible exposure limits to different hazards depends on the degree of risk that is acceptable scientifically and socially, and cost versus benefit analyses are necessary, including the economic impact of such controls. With world opinion varying so widely on what exposures to electromagnetic fields are acceptable it is not surprising that very different criteria have been so far applied in different countries. For instance, Baranski and Czerski (1976) suggest 3 possible approaches to determine *safe* limits. Summarized briefly these are:

1) no demonstrable effects;

2) discernible effects but no change in functional efficiency;

3) exposure results in some stress but only within the limits of normal physiological compensation.

Among the many factors that go into the development of an exposure standard, the selection of a good scientific biological effects data base plays quite obviously the most important role: different scientific approaches have produced different philosophies of protection guidelines and thus different exposure limits (Grandolfo, 1985, 1986).

Very few Western European countries - if any - have a legislation regarding both occupational and population exposure to radiofrequency electromagnetic fields. However, most countries have some form of regulation of the occupational exposure for frequencies above about 10 MHz, either as limits, standards or as guidelines.

The setting of standards for the working environment is always a political decision. Usually the body that issues the standard has a board which consists of representatives of all parties on the labour market, for instance the trade unions and the trade associations. The proposal for a new standard is thus discussed on a risk-benefit basis, where the economical impact is put against the risk of a possible health hazard. The practical aspects of the enforcement of the standard also must be taken into account. Here things like available instruments for measurements of RF-field strengths and frequency, calibration, accuracy, etc. are taken into consideration in the standard setting.

Thus, with this in mind it is understandable that it is not always possible to achieve a standard that for all possible exposure situation would give, for instance, SAR-values below a certain level, but those values would occur so seldom that the hazard involved is estimated to be minimum and acceptable.

In recent years there have been many comprehensive reviews of the scientific basis for setting standards intended for the control of human exposure to non ionizing electromagnetic fields, and here the ground will not be covered again in detail other than to present some general ideas on the debate which has accompanied the recent approval of new national and international safety standards, which in turn proves the need for a new data base as useful input for a definite risk assessment.

These new exposure standards are based on what is known about the frequency-dependent nature of radiofrequency radiation energy deposition in biological systems and the current knowledge of biological effects, and in general provide an added margin of safety over what was previously used.

In this paper an attempt will be made to present a comparative analysis of these existing or recently proposed standards.

INTERNAL DOSIMETRY

Scientists today can look back with some satisfaction over the progress that has been made in the past several years, in developing methods of intercomparing the results of research studies and of relating them to the human situation, to facilitate standards-making.

A principal step forward has been that concerning internal dosimetry, the quantity designated as the specific absorption rate (SAR), expressed in watts per kilogram. The SAR, as recommended by the National Council on Radiation Protection and Measurements (NCRP) is defined as *the time rate at which radiofrequency electromagnetic energy is imparted to an element of mass of a biological body.*

The inherent risks to health from RF/MW exposures are directly linked to the absorption and distribution of energy in the body, and the absorption and distribution are strongly dependent on the size and orientation of the body and the frequency and polarization of the incident radiation (Mitchell, 1985; Durney et al., 1986; Gandhi et al., 1979; Gandhi, 1986; Grandolfo et al., 1983; Polk and Postow, 1986; Michaelson and Lin, 1987).

Both theory and experiment show that RF/MW absorption in prolate spheroid models approaches a maximum value when the long axis of the body is both parallel to the electric field vector and equal to approximately four-tenths of the wavelength of the incident RF/MW field.

This frequency-dependent behaviour is illustrated in Fig. 1 for several human sizes. The average whole-body SAR in W/kg is plotted as a function of radiation frequency in MHz for an incident average power density of 1 W/m^2.

Based on absorption characteristics in the human body, the RF/MW range can be subdivided into four regions (Schwan 1982a; IRPA, 1988), as shown in Fig. 2:

a) the sub-resonance range, less than 30 MHz, where surface absorption dominates for the human trunk, but not for neck and legs, and where energy absorption increases rapidly with frequency;

b) the resonance range, which extends from 30 MHz to about 300 MHz for the whole body, and up to about 400 MHz if partial body resonances, more particularly in the head, are considered. High absorption cross-sections are possible, and exposure limits must be set at lower values to account for worst case situations;

c) the *hot-spot* range, extending from about 400 MHz up to 2 GHz or even to 3 GHz, where significant localized energy absorption can be expected at incident power densities of about 100 W/m^2. Energy absorption decreases when frequency increases and the size of hot spots ranges from several centimeters at 915 MHz to about one centimeter at 3 GHz;

d) the surface absorption range, over about 2 GHz, where the temperature elevation is localized at the surface of the body.

During the past 10 years, this variation of energy absorption with frequency has become the most important parameter considered in setting the latest exposure limits.

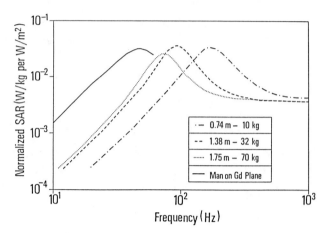

Fig. 1. Specific absorption rate for different size humans.

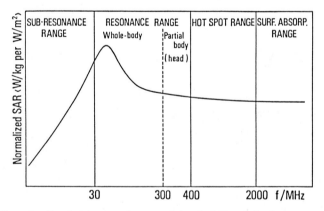

Fig. 2. Variations of normalized SAR with frequency.

WORLDWIDE STANDARDS

In considering standards, it is necessary to keep in mind the essential differences between an *exposure* standard and a *performance* or *emission* standard for a piece of equipment. An exposure standard refers to the maximum level of power density and exposure time for the whole body or for any of its parts, and incorporates a safety factor. An emission standard (or performance standard) refers not to people but to equipment, and specifies the maximum emission (or leakage) from a device at a specified distance. Emission standards are such that human exposure will be at levels considerably below exposure limits.

Subsequent to the publication of the exposure standards of the USSR and the USA most countries that felt the need for radiofrequency exposure standards elected to work to one or other of these standards. Some other countries have adopted, or become involved in the development of standards that, in general, incorporate maximum exposure limits that are intermediate between those of the US and Soviet standards. A review of all national standards in force in 1988 is beyond the scope of this paper, however, it seems useful to present here some recently adopted or proposed standards that in some cases reflect diminishing differences between the various approaches to exposure limits.

Australia

In Australia a standard has been promulgated, effective 31 January, 1985, covering electromagnetic radiation with a frequency ranging from 300 kHz to 300 GHz (SAA, 1985). The standard applies to the exposure of radiation workers due to their employment and the incidental exposure of the general public, but it does not apply, as usual, to patients undergoing medical diagnosis or treatment.

It is intended that the Australian standard will have a limited life and will be reviewed within three years on the basis of evidence gained during its application and from research on the subject which is now world-wide.

In an area in which the possibility of radiofrequency shock and burns exists, the mean squares of the electric and magnetic field strengths, when averaged over any 60-second period, shall not exceed the appropriate occupational limits given in Table 1, except for limited period exposure. The limits valid in an area in which the possibility of

Table 1. Australian occupational limits in an area in which the possibility of RF shocks and burns exists. Values are averaged over any 60-second period

Frequency (MHz)	Mean square electric field $(V/m)^2$	Mean square magnetic field $(A/m)^2$	Mean equivalent plane wave power density (W/m^2)
0.3-9.5	3.77×10^4	0.265	100
>9.5-30	$3.39 \times 10^6 / f^2$	$23.9/f^2$	$9,000/f^2$
>30-3x10^5	3.77×10^3	2.65×10^{-2}	10

f = frequency in MHz

Table 2. Australian occupational limits in an area in which the possibility of RF shocks and burns has been eliminated. Values are averaged over any 60-second period

Frequency (MHz)	Mean square electric field $(V/m)^2$	Mean square magnetic field $(A/m)^2$	Mean equivalent plane wave power density (W/m^2)
0.3-3	3.77×10^5	2.65	1,000
>3-30	$3.39 \times 10^6/f^2$	$2.39/f^2$	$9,000/f^2$
>30-3x10^5	3.77×10^3	2.65×10^{-2}	10

f = frequency in MHz

radiofrequency shock and burns has been eliminated are shown in Table 2. Where non-occupational exposure to an electromagnetic field occurs, the mean squares of the electric or magnetic field strengths shall not exceed one-fifth of the levels specified in Tables 1 and 2. According to the Australian standard, medical surveillance shall be made available to occupationally exposed workers in order to assess the health of individual workers, to help in ensuring initial and continuing compatibility between the health of workers and their conditions of work, and to provide an information base in instances of accidental exposure or occupational disease.

Austria

On December 1, 1986, the Austrian Standards Institute published a draft document containing proposals for the protection of workers and general public against radiofrequency and microwave radiations in the frequency range up to 3000 GHz (near infrared). The document (ONORM, 1986) has no legal backing. As most national standard issues, it has of a voluntary or consensus nature and thus provides only guidelines for exposure. Tables 3 and 4 summarize the levels of root mean square (rms) electric and magnetic field strengths recommended for workers and member of the public, respectively.

Table 3. Austrian occupational exposure limits (rms values) to radiofrequency and microwave electromagnetic fields

Frequency (MHz)	Electric field (V/m)	Magnetic field (A/m)
0.01-0.03	1,500	350
0.03-2	1,500	$7.05706/f^{1.11332}$
2-30	$3,404.61/f^{1.18253}$	$7.05706/f^{1.11332}$
30-3,000	61	0.16
3,000-12,000	$0.570057 f^{0.583647}$	$0.0013881 f^{0.592933}$
12,000-3x10^6	137	0.36

f = frequency in MHz

Table 4. Austrian general public exposure limits (rms values) to radio-frequency and microwave electromagnetic fields

Frequency (MHz)	Electric field (V/m)	Magnetic field (A/m)
0.01-0.03	670	157
0.03-2	670	$3.15601/f^{1.11332}$
2-30	$1522.59/f^{1.18253}$	$3.15601/f^{1.11332}$
30-3,000	27	0.072
3,000-12,000	$0.254937 f^{0.583647}$	$0.62078 \times 10^{-3} f^{0.592933}$
12,000-3x10^6	61	0.16

f = frequency in MHz

Belgium

The only regulation regarding RF radiation in Belgium is given as a general rule regarding protection of man and environment against harmful effects and nuisance caused by non-ionizing radiation, infra- and ultrasound. Issued in 1985, this law is only a general framework that gives the possibility to issue standards for RF fields. However, no such a standard has been enforced so far.

Canada

The Canadian standard was developed following an indepth scientific evaluation of the literature and was proposed (Repacholi, 1978) as a draft so that it could be extensively reviewed. The final standard applies to both occupational exposure and exposure of the general population (HWC, 1979). Table 5 summarizes the occupational exposure limits for whole or partial body exposure to either continuous or modulated electromagnetic radiation in the frequency range 10 MHz-300 GHz. In the same frequency range, an unlimited esposure of general public is allowed up to a power density of 10 W/m².

Table 5. Canadian occupational exposure limits for whole or partial body exposure continuous or intermittent radiation from 10 MHz-300 GHz (averaged over 1 h)

Frequency (MHz)	Exposure limits
10-1,000	10 W/m²
	60 V/m
	0.16 A/m
1,000-300,000	50 W/m²
	140 V/m
	0.36 A/m

The Canadian government has proposed revisions in its radiofrequency and microwave exposure standards that will make it much stricter than before (Stuchly, 1987). It is following the lead of the U.S. National Council on Radiation Protection and Measurements (NCRP) and is planning to reduce allowable public exposures to RF/MW energy by a factor of five.

The responsible agency is the Canadian Bureau of Radiation and Medical Devices. In Tables 6 and 7 the complete set of proposed exposure limits for occupational and general populations is given.

Table 6. Canadian proposed exposure limits (rms values) for radiofrequency and microwave (Occupational)

Frequency (MHz)	Electric field (V/m)	Magnetic field (A/m)	Power density (W/m²)
0.01-1.2	600	4	-
1.2-3	600	4.8/f	-
3-30	*1,800/f or $3,120/f^{1.5}$	4.8/f	-
30-100	*60 or 20	0.16	-
100-300	*60 or 0.2f	0.16	10
300-1,500	$3.45f^{0.5}$	$0.0093f^{0.5}$	0.032f
1,500-300,000	140	0.36	50

f = frequency in MHz
* The lower limits apply only when the exposed person is separated less than 0.1 m from electrical ground; in all other cases the higher limits apply.

Table 7. Canadian proposed exposure limits (rms values) for radiofrequency and microwave (General population)

Frequency (MHz)	Electric field (V/m)	Magnetic field (A/m)	Power density (W/m²)
0.01-1.2	280	1.8	-
1.2-3	280	2.1/f	-
3-30	*840/f or $1,600/f^{1.5}$	2.1/f	-
30-100	*28 or 10	0.07	-
100-300	*28 or 0.1f	0.07	2
300-1,500	$1.61f^{0.5}$	$0.004f^{0.5}$	$6.44 \times 10^{-3}f$
1,500-300,000	60	0.16	10

f = frequency in MHz
* The lower limits apply only when the exposed person is separated less than 0.1 m from elctrical ground; in all other cases the higher limits apply.

Czechoslovakia

According to Marha (1970), the available biological data identify 10 µW/cm² average power density for pulsed microwaves and 25 µW/cm² for continuous waves over 8 hour work day as an exposure level (EL) for workers, taking a 10-fold safety factor into account.

On this basis, guide numbers for establishing ELs for a given duration of exposure were derived. Thus in the microwave range (300 MHz to 300 GHz) the ELs for a given duration of exposure during the working day are calculated from the formulas:

EL = 2/t (continuous waves)

EL = 0.8/t (pulsed waves)

where t is the time of exposure in hours, and ELs are expressed in W/m².

The above may be interpreted that 0.8 W/m² or 2 W/m² for one hour exposure to pulsed or continuous microwaves are considered ceiling levels.

To derive public ELs ten times lower values and a duration of exposure of 24 hours were adopted. Thus, public ELs are derived from the formulas:

EL = 0.6/t (continuous waves)

EL = 0.24/t (pulsed waves)

where t is the time of exposure in hours, and ELs are expressed in W/m².

The above may be interpreted that 0.24 W/m² or 0.6 W/m² for one hour exposure to pulsed or continuous microwaves are considered ceiling levels, and 10 mW/m² or 25 mW/m² over 24 hours are the basis for deriving public ELs.

A similar approach was taken to ELs at lower frequencies, as presented in Tables 8 and 9. The values given in Tables 8 and 9 are based on 1970 regulations (Czerski, 1985).

According to Musil (1983), changes in the Czechoslovakian ELs were proposed, as shown in parts B of Tables 8 and 9. No information on the final outcome, acceptance or nonacceptance of the changes is presently available to the present authors.

Two features of the 1970 Czechoslovakian regulations deserve to be stressed (Czerski, 1985). One is the distinction between ELs for exposure to continuous or pulsed microwaves, and the adoption of more restrictive values for the latter. The second is the time-dependence of the ELs, expressed by *guide numbers* derived from ELs for the duration of 8 hours (working day, occupational) or 24 hours (public ELs).

There is currently the will to reach the unification of electromagnetic fields safety regulations in Comecon countries (Council for Mutual Economic Assistance) following a proposal elaborated by Soviet experts. The Soviet proposal, however, is still a subject for criticism.

Table 8. Czechoslovakian radiofrequency occupational exposure limits (A), and proposed changes (B)

Frequency (MHz)	Workday (8 h)	For shorter periods	Comments
A.			
0.03-30	50 V/m	(400/t) V/m	
30-300	10 V/m	(80/t) V/m	
300-300,000	0.25 W/m²	(2/t) W/m²	Continuous waves
	0.1 W/m²	(0.8/t) W/m²	Pulsed waves
B.			
0.03-10	70 V/m		No details on
10-70	(700/f) V/m		proposed ELs for
70-300	10 V/m		shorter periods
300-500	0.25 W/m²		available for all
500-300,000	0.25 W/m²		frequency ranges

t = duration of exposure in hours; f = frequency in megahertz

Table 9. Czechoslovakian radiofrequency public exposure limits (A), and proposed changes (B)

Frequency (MHz)	24-h continuous exposure	For shorter periods	Comments
A.			
0.03-30	5 V/m	(120/t) V/m	
30-300	1 V/m	(24/t) V/m	
300-300,000	0.025 W/m²	(0.6/t) W/m²	Continuous waves
	0.01 W/m²	(0.24/t) W/m²	Pulsed waves
B.			
0.03-10 MHz	14 V/m		No details on
10-70 MHz	2 V/m		proposed ELs for
70-300 MHz	2 V/m		shorter periods
300-500	$(0.1f-20) \times 10^{-3}$ W/m²		available for all
500-300,000	0.03 W/m²		frequency ranges

t = duration of exposure in hours; f = frequency in megahertz

Federal Republic of Germany

In August 1986 a proposal of DIN VDE standard was published in West Germany (DIN VDE, 1986). The proposal covers the frequency range from 0 Hz to 3,000 GHz. The limits are claimed to be set in accordance with today's knowledge to avoid hazards from exposure to EM fields for healthy persons.

The standard applies to the general public as well as to workers. Only those effects are accounted for that may have an acute influence on health such as stimulation of receptors and individual cells, influences of the heart activity, etc.

Effects, that are only associated with perception of electric or magnetic fields such as visible perception within the magnetic field (magnetic phosphenes), surface effects such as hair motion, microshocks, spark discharges, perception of short-time discharge or continuous currents have not been accounted for.

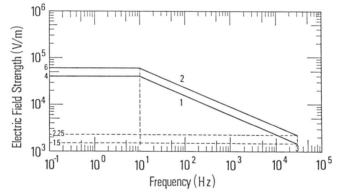

Fig. 3. Proposed West German limits for the electric field strength for 0 Hz < f < 30 kHz for whole body exposure. Curve 1 shows the allowed rms-values and curve 2 the ceiling values.

For frequencies above 30 kHz the limits are set on a purely thermal basis, and such that a sufficient margin of safety is kept to SAR-values of 4 W/kg, where unwanted effects occur according to animal experiments. In brief the proposal is presented in Figs. 3-6. For further details see Rozzell (1985).

To be noted is that special values applies to pacemaker wearers in the frequency range up to 30 MHz, and for higher frequencies limits are under preparation.

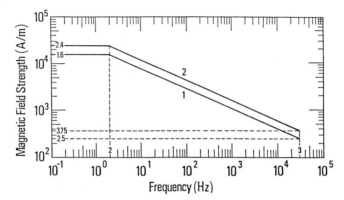

Fig. 4. Proposed West German limits for the magnetic field strength for 0 Hz < f < 30 kHz for whole body exposure. Curve 1 shows the allowed rms-values and curve 2 the ceiling values.

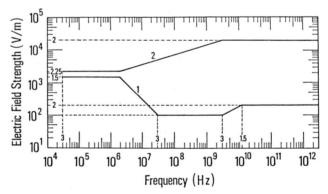

Fig. 5. Proposed West German limits for the electric field strength for 30 kHz < f < 3,000 GHz for whole body exposure. Curve 1 shows the allowed time-averaged values (6 min and more) and curve 2 the ceiling values.

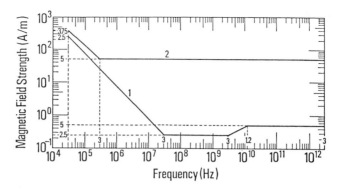

Fig. 6. Proposed West German limits for the magnetic field strength for 30 kHz < f < 3,000 GHz for whole body exposure. Curve 1 shows the allowed time-averaged values (6 min and more) and curve 2 the ceiling values.

Finland

In June 1985 the Finnish government issued a law regarding stray fields from high frequency (10 MHz-100 MHz) apparatuses. The basic exposure limits are 60 V/m and 0.2 A/m for a typical working day. These values are rms field strengths determined in a 6-minute period or the mean from five cycles. Ceiling values never to be exceeded are also given; 300 V/m and 0.8 A/m, respectively. Measurements should be taken immediately after installation of the equipment and the protocol should be sent to the National Board of Occupational Safety and Health. New measurements should be taken at least every third year or whenever a change has been made on the machine that might influence the stray fields. The accuracy in the measurements should be within 1 dB.

For exposure times shorter than 1 h of the working day the highest permissible values are obtained by the formula:

$t = k/P$

where $k = 36,000$ Ws/m^2, t the total time in seconds, and P is the equivalent power density in watts per square meters.

These regulations are valid for all new equipment installed after January 1st, 1986, and from January 1st, 1989, they are also valid for older equipment.

According to the Finnish Centre for Radiation and Nuclear Safety no other official exposure standard or even proposal exists for RF radiation. A preliminary paper regarding occupational exposure and general public has been presented, which follows in principle the IRPA guidelines (Jokela, 1988).

France

In France no rules are given for RF fields and occupational and public exposure. The Institut National de Recherche et de Securite' has published a series of paper regarding the hazards from RF exposure and how to reduce stray fields (INRS, 1978, 1982, 1983, 1985). Their recommendations regarding limits follow the ACGIH values closely.

Italy

In Italy, a national Committee charged by the Minister of Health published a draft proposal for exposure limits in the frequency range of 300 kHz-300 GHz. The draft, which has not yet received final approval from the Italian Parliament, indicates the limits of exposure to electromagnetic fields for workers and members of the public. These levels refer to both total and partial body exposures, caused by one or more sources of continuous or modulated wave emissions.

For the frequency range between 300 kHz and 3 MHz the occupational exposure levels, averaged over any 1/10 hour period, must be lower than:

- rms electrical field strength   140 V/m
- rms magnetic field strength   0.36 A/m
- mean power density   50 W/m^2

and correspondingly, during the same period of time they must not exceed:

- rms electrical field strength   300 V/m
- rms magnetic field strength   0.83 A/m
- mean power density   250 W/m^2

For the frequency range between 3 MHz and 300 GHz the occupational exposure levels, averaged over any 1/10 hour period, must be lower than:

- rms electrical field strength   60 V/m
- rms magnetic field rength   0.17 A/m
- mean power density   10 W/m^2

and correspondingly, during the same period of time they must not exceed:

- rms electrical field strength   200 V/m
- rms magnetic field strength   0.50 A/m
- mean power density   100 W/m^2

According to this draft proposal medical surveillance is compulsory for exposed workers, and is guided by the principles which govern industrial medicine and is entrusted to occupational health specialists. The recommended examinations are quite similar to those included in a NIOSH document (Glaser, 1980).

For the frequency range between 300 kHz and 3 MHz, the general public exposure exposure levels must not exceed the maximum values of:

- electrical field strength   45 V/m
- magnetic field strength   0.11 A/m
- mean power density   5 W/m^2

For the frequency range between 3 MHz and 300 GHz the general public exposure levels must not exceed the maximum values of:

- electrical field strength   20 V/m
- magnetic field strength   0.05 A/m
- mean power density   1 W/m²

Presently, no nationwide consensus has been reached on this proposal and today these limits have a legal backing only in an Italian region (Piemonte).

A Task Group has been recently formed to propose changes in the recommended exposure limits, in order to take into account the response to the former proposal and the further development of the understanding of the biological effects of radiofrequency and microwave radiation.

Norway

The State Institute of Radiation Hygiene is the body that issues the standards in Norway, and so far they only have a proposal since 1982 (Saxebol, 1982) for administrative norms.

The norm concerns whole body radiation for up to 8 h per day. The values for the workplace are given in Table 10.

For the general public a safety factor of ten is used for the equivalent power density values.

The values are mean values over 1 h, and the ceiling power density value never to be exceeded for 6 MHz - 300 GHz is 250 W/m².

The recommendation for general public exposure is to keep that no higher than a factor of ten below the above limits for the equivalent power density.

These administrative norms are based upon technical, economical and medical assessments, and are not legally enforced.

Table 10. Proposed Norwegian standard for radiofrequency and microwaves

Frequency (MHz)	Mean square electric field $(V/m)^2$	Mean square magnetic field $(A/m)^2$	Mean equivalent plane wave power density (W/m^2)
1-6	10^5	0.625	250
6-30	$36 \times 10^5/f^2$	$22.5/f^2$	$9{,}000/f^2$
30-1,000	4,000	0.025	10
1,000-300,000	20,000	0.125	50

f = frequency in MHz

Poland

In 1972, Poland departed from the Soviet standard on the basis of an extensive 15-year study of a large number of civilian and military personnel engaged in work from one to more than 15 years under exposure conditions conforming to the Polish standards quoted below. During this study, no instances of irreversible damage or disturbances caused by exposure microwave radiation were encountered. Any disturbances and deviations from normal were those of a functional nature. Medical examinations conducted several months after removal from occupational exposure to microwave radiation indicated a reversal of disturbances. A group of selected subjects, who were examined in detail during a five-year period of observation, were found to be healthy (Czerski and Piotrowski, 1972) even in instances where permissible radiation levels had been exceeded. The exposed group presented a better state of health than a group of corresponding age, length of employment, and living conditions, not occupationally exposed to microwave radiation.

Under the Polish standard, in the frequency range from 300 MHz to 300 GHz a distinction is made between stationary fields, continuous irradiation at a given point, such as a work station or personnel position, and interrupted irradiation at such a point, designated as nonstationary fields. On the basis of these principles, they have differentiated safe, intermediate, hazardous, and danger zones, where the boundaries of the individual zones are determined by measuring average power density levels in the bands from 300 MHz to 300 GHz in watts per square metre (Czerski, 1985).

Stationary fields were defined as fields generated by equipment with stationary or mobile antennas or beams, where an index C is greater than 0.1; nonstationary fields were defined by the index C smaller than 0.1 and the irradiation cycle rate greater than 0.02. The index C is defined by the formula:

$$C = t/T$$

where t is the irradiation time, i.e. the period of time during which the power density of the moving beam is greater or equal to one half of the maximum average power density (for one cycle of the movement of the beam), and T is the duration of one cycle of the movement of the beam.

The following zone boundaries are proposed for stationary fields:

1) Safe zone - the highest level of mean power density shall not exceed 0.1 W/m^2; human exposure is unrestricted.

2) Intermediate zone - the boundary values of radiation level shall be 0.1 W/m^2 at the lower boundary and 2 W/m^2 at the upper boundary. Occupational exposure is allowed during a whole working day.

3) Hazardous zone - the lower and upper boundary levels shall be 2 W/m^2 and 100 W/m^2. The occupational exposure time is determined by a formula.

4) Danger zone - exposure levels greater than 100 W/m^2. Human exposure is forbidden.

For exposure to non-stationary fields, i.e. intermittent exposure, the following values were adopted:

1) Safe zone - mean power density not to exceed 1 W/m^2; human exposure is unrestricted.

2) Intermediate zone - minimum value 1 W/m², upper limit 10 W/m²; occupational exposure allowed during a whole working day.

3) Hazardous zone - minimum value 10 W/m², upper limit 100 W/m². The occupational exposure time is determined by a formula.

4) Danger zone - mean power density in excess of 100 W/m²; human exposure forbidden.

The Polish law names the bodies to be responsible for health surveillance, supervision of working conditions, and the manner of carrying out the measurements (in principle, every 3 years, and after changes in equipment or its diplacement).

Permissible times for a worker in a hazardous zone are determined by the formulas:

Stationary field $\quad t = 32/p^2$

Nonstationary field $\quad t = 800/p^2$

where t = time in hours, and p = average power density level in watts per square metre.

The occupational exposure standard for the frequency range from 100 kHz to 300 MHz is summarized in Table 11.

The public exposure limits make a distinction between unrestricted occupancy (habitation) and occasional short-time human presence in the field. In the frequency range from 0.1 to 10 MHz unlimited occupancy is allowed at field strengths below 5 V/m, and short-term presence of the public (less than a day) is allowed up to 20 V/m. The corresponding values for frequencies between 10 and 300 MHz are 2 and 7 V/m, and for frequencies between 300 MHz and 300 GHz are 0.25 and 1 W/m². Public exposure to fields greater than 20 V/m, 7 V/m and 1 W/m² for respective frequency ranges is prohibited.

Table 11. Polish occupational electric field exposure levels from the frequency range 100 kHz-300 MHz

Zones	Frequency	
	100 kHz-10 MHz	10 MHz-300 MHz
Safe	< 20 V/m	< 7 V/m
Intermediate (8 or 10 hours)	20-70 V/m	7-20 V/m
Hazardous	70-1,000 V/m (t = 560/E)	20-300 V/m (t = 3,200/E²)
Danger	> 1,000 V/m	> 300 V/m

E = electric field strength in V/m
t = duration of exposure in hours

Sweden

In Sweden the limits are set by the National Board of Occupational Safety and Health (NBOSH), and for the radiofrequency radiation in workplaces a standard has been in effect since 1977. The permissible values were given as 6-minute time-averaged values and ceiling rms values over 1 s. For frequencies from 10 MHz to 300 MHz the long term value was 50 W/m² and from 300 MHz to 300 GHz 10 W/m². The ceiling value was 250 W/m² for the whole frequency range.

A new set of values have now been issued by the Board, valid from January 1st, 1988, and covering the frequency range down to 3 MHz. The limits are based on thermal considerations and the documentation behind the IRPA interim guidelines (IRPA, 1984) and the ANSI C95.1-1982 standard have been used as starting points for the standard setting. In contrast to the previous Swedish regulation giving the values in terms of power density, the electric and the magnetic field strengths are now the fundamental quantities being restricted. The recent finding of whole body resonance at about 30-40 MHz for persons in contact with high frequency electric ground has been taken into account as a lowering of the values in these cases (Gronhaug and Busmundrud, 1982; Hill, 1984). In Table 12 the new values are given. The field strength values are for positions where the operator can be, but should be measured as undisturbed fields. The values are given either as time-averaged over any 6-minute period during the working day or as ceiling values never to be exceeded. For pulsed fields these are time-averaged over 1 s. When the worker can be closer than 10 cm to what can be considered as high frequency ground, the values in Table 12 should be replaced in the frequency range 3-60 MHz by the values shown in Table 13.

An exclusion paragraph has also been included about the mobile radio equipment operating with power less than 7 W. In the frequency range up to 1 GHz the radiation emitted from an extended antenna is excluded from the limits given above. If there is a possibility that exposure in excess of the standard can occur, measurements are required to clarify the exposure situations. Appropriate measures should be undertaken to reduce the exposure below the allowed values. Under the new law, the manufacturers of high frequency apparatuses are requested to provide users with safety instructions for the operating of the equipment in such a way that the exposure is less than the given limits.

Persons with metal implant should avoid exposure to EM fields. Pacemaker wearers are requested to consult with their physician before starting to work in places were there is a possibility of EM exposure.

Table 12. Swedish occupational exposure limits as given by the National Board of Occupational Safety and Health

Frequency (MHz)	Time-averaged values E (V/m)	H (A/m)	Ceiling values E (V/m)	H (V/m)
3-30	140	0.40	300	0.8
30-300	60	0.16	300	0.8
300-300,000	60	-	300	-

Table 13. Swedish occupational exposure limits when workers can be closer than 10 cm to the high frequency ground

Frequency (MHz)	Time-averaged values E(V/m)	H(A/m)	Ceiling values E(V/m)	H(A/m)
3-30	47	0.13	100	0.27
30-60	20	0.05	100	0.27

The reason for not using the frequency dependent limits, as it is done by IRPA and ANSI for the high frequency band, is that it would be very cumbersome for the labour inspectorate to include also frequency measurements in the case of, for instance, glue dryers and RF sealers besides field measurements. These machines are often not frequency stable during the operation cycle and, thus, a value would have to be picked for each machine and different limits given. Furthermore, it would be very difficult for the men operating the different machine to make the distinction between these different limits. In view of this it was thought better to have limits constant over a wide frequency range than a slope in frequency. The constraints imposed on for instance 13 MHz machines by this is more than compensated for by the gain in simplicity at 27 MHz.

Guidelines for the general public exposure in Sweden are issued by the National Institute of Radiation Protection. In a recommendation issued in 1978 it is said that the power density should be kept below 10 W/m^2, but if possible the level of 1 W/m^2 should be applied for long term exposure. The Institute has declared that very shortly new guidelines will be issued.

The NBOSH has also an ordinance (AFS 1979:6) regarding the emission from microwave ovens used professionally. The measurements should be performed at 5 cm distance from the surface of the oven, and the load in the oven is prescribed to be 275 ml of a 1% sodiumchloride solution. The allowed leakage is no more than 20 W/m^2 at delivery. A periodical check should be made at least every third year. The allowed leakage is then 50 W/m^2 at 5 cm.

United Kingdom

In 1982, the National Radiological Protection Board (NRPB) published a proposal regarding permissible limits for exposure to RF fields (NRPB, 1982; Allen and Harlen, 1983). The principal basis for the Board's proposal was that exposure at levels likely to cause harmful effects should be prevented. For occupational exposures, levels at which perceptible but harmless effects occur are permissible but should be avoided if possible. Perceptible but harmless effects should be avoided for the general public too. For microwave and radiofrequency radiation the Board accepts whole body SAR less than 0.4 W/kg.

Now NRPB has prepared a consultative document based on the earlier proposals and comments received on them (NRPB, 1986). The draft consists of a set of basic limitations on electric currents and current densities in the body which apply to frequencies roughly below 500 kHz, and above this frequency a set of restrictions on the rate of power dissipation in the body.

In Tables 14 and 15 the guidelines for whole body occupational exposure are given. It can be noted that the proposal includes even static fields. Also to be noted is that the ratio of E/H, i.e. space impedance, varies from 377 Ω at microwave frequencies to 20 Ω at 50 kHz.

Guidelines for public exposure are also proposed, and at frequencies below 1 MHz the limits are set to avoid electric shocks from large ungrounded metal objects in electric fields. The guidelines also take into account the theoretically different absorption characteristics of small children compared to adult from 100 MHz to 1.5 GHz. The values shown in Tables 16 and 17 are about a factor 2-3 lower than for occupational exposure, and for further details see NRPB (1986).

Recently a special guidance note regarding the RF dielectric heating equipment was published by the UK Health and Safety Executive (HSE, 1986). The Guidance Note explains how RF heating equipment works, discusses the hazards involved and gives advice on the precautions that should be observed to ensure safe use. The note also contains guidance for the manufacturers suppliers on how to comply with the requirements in the Health and Safety at Work Act, 1974.

Table 14. NRPB guideline levels for whole body occupational exposure to frequencies in the range 50 kHz to 30 MHz (rms values) for an average total period not exceeding 2 hours per day

Frequency (MHz)	Electric field (V/m)	Magnetic field (A/m)
0.05-0.3	2,000	5/f
0.3-10	600/f	5/f
10-30	60	5/f

f = frequency in MHz

Table 15. NRPB guideline levels for whole body occupational exposure to frequencies in the range 30 MHz to 300 GHz (rms values) for an average total period not exceeding 2 hours per day

Frequency (MHz)	Power density (W/m^2)	Electric field (V/m)	Magnetic field (A/m)
30-100	10	60	0.16
100-500	f/10	$6f^{0.5}$	$0.016f^{0.5}$
500-300,000	50	135	0.36

f = frequency in MHz

Table 16. NRPB guideline levels for areas of public access for frequencies in the range 50 kHz to 30 MHz (rms values) for an average total period not exceeding 5 hours per day

Frequency (kHz)	Electric field (V/m)	Magnetic field (A/m)
50-365	800	$2.0 \times 10^6/f$
365-475	800	5.5
475-580	$380/f$	5.5
580-10,000	$380/f$	$3.2/f$
10,000-30,000	38	$3.2/f$

f = frequency in MHz

Table 17. NRPB guideline levels for areas of public access for frequencies in the range 30 MHz to 300 GHz (rms values) for an average total period not exceeding 5 hours per day

Frequency (MHz)	Power density (W/m²)	Electric field (V/m)	Magnetic field (A/m)
30-300	4	38	0.10
300-1,500	$f/75$	$2.2f^{0.5}$	$0.006f^{0.5}$
1,500-300,000	20	85	0.23

f = frequency in MHz

The occupational exposure levels are determined from the usual thermal considerations, plane wave (far field) characteristics and worst case interactions. The levels given are such that the average whole-body SAR is below 0.4 W/kg and the peak rate, 6-minute average, in any volume of 1 cm³ or less is below 4 W/kg. The basic levels referred to are the ones given in the NRPB (1982) proposal, i.e. in the range 3-30 MHz power density $9,000/f^2$, where f is the frequency in MHz, electric field strength $1,800/f$ and magnetic field strength $5.0/f$. In the range 30-100 MHz the values are 10 W/m², 60 V/m and 0.16 A/m, respectively.

Research in the UK and elsewhere has shown that RF dielectric heater applicators produce fields of limited aperture at the operator positions. The RF energy coupled to the operator is much lower than that of uniform extended field. Until further research has been completed and a UK National Standard has been agreed the exposure should not exceed 10 times the equivalent power density levels given above. At 27 MHz this comes to: 123 W/m², 211 V/m and 0.59 A/m. If the RF emission is less than continuous, these levels should be adjusted for the duty cycle of the machine.

USA

The limit of 100 W/m², which is the current U.S. federal standard (29 CFR 1910.97), was first proposed by Schwan in 1953 after his pioneering work and is based mainly on physiological considerations, in particular of thermal load and heat balance. The maximum rate of working that can be sustained over a period of hours requires the body to dissipate about 750 W. Total absorption of 100 W/m² over the cross-sectional body area of 0.7 m² would result in an additional heat load of 70 W which is small compared with that caused by even the lightest manual activity, and is in fact less than the resting metabolic rate. The 100 W/m² level is also a factor of at least 10 below the exposures that had been found to be needed to cause injury to the testes or the eye, the two organs widely regarded as being most sensitive to thermal injury. The acknowledged assumption in this standard is that heating of body tissues is the most if not the only significant consequence of absorbing radiofrequency energy. The US standard offers the advice that lower levels should be considered if there are other heat stresses, while higher levels might be appropriate under conditions of intense cold.

The American National Standards Institute (ANSI) has recently completed a review of literature (ANSI, 1982) and screened only those reports that produced positive findings, were reproducible, and supplied adequate dosimetric information. Evaluations of these selected reports, in terms of whether or not the biological effects constituted a hazard to health, were made.

Behaviour in experimental animals was found to be the most sensitive indicator of an adverse health effect (e.g. convulsion activity, work stoppage, work decrement, decreased endurance, perception of the exposing field and aversion behaviour). On the basis of their review, ANSI concluded that acute (less than 1 hour) exposure to electromagnetic energy that is deposited in the whole body at an average specific absorption rate (SAR) of less than 4 W/kg does not produce an adverse health effect in experimental animals. However, because prolonged exposure (days and weeks) may cause damage, a tenfold reduction in the permissible SAR (i.e., 0.4 W/kg) was invoked.

The ANSI proposals, shown in Table 18, apply to both occupational and general population exposures and cover the frequency range 300 kHz-100 GHz. Exposures are to be averaged over 6 minutes, and levels in excess of those in Table 18 are allowed if it can be demonstrated that the mean specific energy-absorption rate is less than 0.4 W/kg and the peak rate in any part of the body does not exceed 8 W/kg. Emissions from devices radiating less than 7 W and at frequencies below 1 GHz are exempt. Above 1 GHz devices will be treated on a case by case basis.

Table 18. ANSI proposals for limiting population and workers exposures to radiofrequency and microwave radiations

Frequency range (MHz)	Power density (W/m²)	Electric field strength (V/m)	Magnetic field strength (A/m)
0.3-3	1,000	632	1.58
3-30	$9,000/f^2$	$1,897/f$	$4.74/f$
30-300	10	63.2	0.158
300-1,500	$f/30$	$3.65f^{0.5}$	$9.12 \times 10^{-3} f^{0.5}$
1,500-100,000	50	141.4	0.353

f = frequency in MHz

Exposures slightly in excess of the proposed limits are not expected to be harmful but are regarded as undesirable. In uncontrolled situations it is suggested that exposure should be as low as reasonably achievable (ALARA principle).

Recent work, however, has pointed to several potential problems with this guideline. They pertain to large RF-induced currents and the commensurately high local SARs for some regions of the body, and contact hazards for commonly-encountered ungrounded objects in ANSI-recommended E-fields for the frequency band 300 kHz-62.5 MHz (Gandhi, 1986). Also, the ANSI-recommended guideline of 50 W/m^2 for the millimeter-wave band (frequency greater than 30 GHz) may be close to the power densities that are likely to cause sensations of *very warm to hot* for whole-body exposures (Gandhi, 1986).

It is interesting to note that for several years the Lawrence Livermore National Laboratory has been developing (Counts, 1982) a program to monitor radiofrequency and microwave radiation sources. The Laboratory as a result of the literature decided to adopt exposure limits more conservative than the ANSI standard; believing that not enough is known about lower frequencies in the kilohertz range, they elected to reduce their exposure standard in the frequency range 10 kHz-3 MHz by a factor of 10 from the ANSI.

In 1988, the American Conference of Governmental Industrial Hygienists (ACGIH) issued Threshold Limit Values (TLVs) referring to radiofrequency and microwave radiation in the frequency range from 10 kHz to 300 GHz (ACGIH, 1988) and representing conditions under which it is believed workers may be repeatedly exposed without adverse health effects. The TLVs shown in Table 19 have been selected to limit SAR to 0.4 W/kg in any 6-minute period for frequencies higher than 3 MHz. Between 10 kHz and 3 MHz SAR is still limited to 0.4 W/kg, but a plateau at 1 kW/m^2 was set to protect against shock and burn hazards.

According to ACGIH these values should be used as guides in the evaluation and control of exposure to radiofrequency/microwave radiation, and should not be regarded as a fine line between safe and dangerous levels. All exposures should be limited according to the ALARA principle given the current state of knowledge on human effects, particularly non-thermal effects. Anyway the TLVs in Table 19 may be exceeded according to ACGIH if the exposure conditions can be demonstrated to produce a SAR of less than 0.4 W/kg as averaged over the whole body and spatial peak values less than 8.0 W/kg as averaged over any 1.0 gram of tissue.

Table 19. ACGIH limit values for radiofrequency and microwave radiation

Frequency (MHz)	Power density (W/m^2)	Electric field strength (V/m)	Magnetic field strength (A/m)
0.01-3	1,000	614	1.63
3-30	9,000/f^2	1,842/f	4.89/f
30-100	10	61.4	0.164
100-1,000	f/10	6.14f$^{0.5}$	0.0163f$^{0.5}$
1,000-300,000	100	194	0.514

f = frequency in MHz

Taking into account that at frequencies below 30 MHz ungrounded objects such as vehicles, fences, etc., can strongly couple to RF fields, for field strengths near the TLV, shock and burn hazards can exist. Care should be taken to eliminate ungrounded objects, to ground such objects, or use insulated gloves when ungrounded objects must be handled. Finally, all exposures to pulsed fields, whatever the frequency, should be limited to a maximum peak electric field strength of 100 kV/m.

Unlike the ANSI standards, the TLVs cover the added frequency range from 10 to 300 kHz and from 100 to 300 GHz. Because the TLVs are to be applied in occupational settings, they assume that no children will be in the workplace. This assumption allows an average incident power density of 100 W/m² at frequencies greater than 1 GHz, while maintaining the same 0.4 W/kg whole-body absorption limit. The ACGIH TLVs are established as safety guidelines for the workplace. They are intended for use in the practice of industrial hygiene and should be interpreted and applied only by a person trained in this discipline.

In May 1978, the National Institute for Occupational Safety and Health (NIOSH) formally initiated the development of a criteria document (Glaser, 1980) and recommended standard for occupational exposure to radiofrequency and microwave radiation to protect workers from these radiations. Table 20 summarizes the proposed occupational exposure limits recommended in the NIOSH criteria document. In addition to the proposed occupational exposure levels, the NIOSH criteria document makes specific recommendations in the areas of monitoring of occupational exposure limits and of medical surveillance. According to NIOSH recommendations, medical surveillance shall be made available to personnel at risk of exposure to radiofrequency and microwave energy. Thorough medical and work histories and physical examinations are suggested despite the lack of well-defined health effects correlated with exposure of humans to RF and microwave energy. Inclusion or exclusion of any or all of the tests recommended as part of the examination shall be at the discretion of the responsible physician.

Procedures suggested for the medical examination include the following:

1) Laboratory examinations, including urine analysis, hematocrit, white blood cell (WBC) count, differential count, and analysis of serum for total protein, blood urea nitrogen (BUN), glucose, albumin, globulin, tetraiodothyronine (T_4), electrolytes, triglycerides, cholesterol, and free fatty acids.

2) Evaluation of cardiovascular function, including an electrocardiogram (ECG).

3) Evaluation of neurologic function, including an electroencephalogram (EEG). An emotional and behavioral profile shall be complied with attention to such factors as weakness, headache, memory impairment, inattention, insomnia, and irritability.

4) Examination of the skin and eyes for evidence of significant exposure to radiofrequency and microwave energy, i.e. erythema or burns of the skin, corneal and lenticular opacities, and conjunctival and corneal injections.

Preplacement examinations shall be made available to all new employees. The examination shall consist of the medical evaluation detailed above and also shall include comprehensive medical and work histories, with particular attention given to previous exposure to ionizing and non-ionizing radiation.

Table 20. NIOSH recommended occupational exposure limits (as averaged over any 6-minute period)

Frequency range (MHz)	Power density (W/m^2)	Electric field strength (V/m)	Magnetic field strength (A/m)
0.3-2	250	307	0.814
2-10	1,000/f^2	614/f	1.628/f
10-400	10	61.4	0.163
400-2,000	f/40	3.07f$^{0.5}$	8.14x10^{-3}f$^{0.5}$
2,000-300,000	50	137.3	0.364

f = frequency in MHz

Annual medical examinations consisting of the procedures described above shall be made available to all workers exposed to radiofrequency and microwave energy at field strengths above the action level. Work histories shall be updated at this time. Examination of employees who are exposed at field strengths at or below the action level is suggested.

Exposure above the permissible occupational exposure levels shall be followed within 3-7 days by an examination consisting of the evaluation suggested above. A follow-up examination is suggested within 1-2 months post exposure but shall be performed at the discretion of the responsible physician.

Determination that exposure above the occupational exposure limits has occurred shall be performed according to the protocol described in the appropriate section. Additional diagnostic procedures that may be useful in assessing the severity of the effects from such exposure are briefly outlined in the draft criteria document.

Medical records including health and work histories shall be maintained for a period of years after employment ends for all persons occupationally exposed to radiofrequency and microwave energy.

The OSHA standard, adopted in 1972, applies to employees in the private sector. An addendum, adopted in 1975, applies to work conditions particularly in the telecommunications industry. OSHA standards are mandatory for federal employees including the military. Maximum permissible exposure limit was 100 W/m^2, for durations greater than 6 min, over the frequency range 10 MHz-100 GHz. The OSHA standard, however, has been challenged as being unenforceable.

It is interesting to note that the United State Environmental Protection Agency (USEPA) has delayed the release of its proposed radiofrequency/microwave radiation exposure guidance, which was scheduled for June, 1984.

In the first official statement to indicate rule making may be abandoned, EPA announced in early June 1984 that the agency is considering a number of options for the guidance, including putting it on ice. Meanwhile, the Office of Research and Development's report on RF/MW bioeffects has received final approval from EPA's Scientific Advisory Board. The document will serve as the basis for the guidance (Elder and Cahill, 1984), if any.

A basic criterion of USSR protection standards is that they should be set at levels where there is no change in the state of health or well being of exposed individuals. In practice, this is taken to be at levels well below those at which there are any perceptible effects of exposure.

Petrov and Subbota (1972) in a review of Soviet work reported a threshold at a power density of 10 W/m^2 for functional disturbances in laboratory animals and people exposed for one hour to 3 GHz radiation. Applying a safety factor of 10 and assuming a reciprocal relationship between exposure and duration led to an overall reduction by a factor of 100 for 10 hour exposures, so that the occupational exposure limit became 0.1 W/m^2. A population limit for continuous exposure, a factor of two less than this, i.e., 0.05 W/m^2, came into force some years later. The occupational limit was increased to 0.25 W/m^2 in January 1982, but the higher limits for shorter exposure were not changed. These are 1 W/m^2 for exposure durations of less than 2 hours and 10 W/m^2 for exposure durations of less than 20 minutes.

There have been reports from the USSR that the exposure of workers to microwaves at levels between about 0.1 W/m^2 and a few W/m^2 has led to *functional disturbances in the nervous system with a predominantly asthenic condition,* including depression, impairment of memory, inability to make decisions, etc. It is claimed that these symptoms are reversible when the exposure ceases and it is noteworthy that they are nonspecific to microwave exposure. There have been similar reports from other eastern European countries but careful studies have failed to confirm these claims.

Detailed data on radiofrequency and microwave occupational exposure limits in USSR (USSR, 1976) are given in Table 21.

The USSR standard which specifies public exposure levels (Czerski, 1985) is shown in Table 22. It is interesting to note that the subdivision of RF radiation into ranges is different in this standard from those established in the occupational one. No public exposure limits are specified for the magnetic field strength.

It appears, however, that to a large degree, the apparent differences in US and USSR standards are based not on actual factual information but on differences in basic philosophy. These differences appear in the areas of industrial hygiene.

The basic industrial hygiene philosophy of the USSR (Magnuson, 1964) can be summarized as follows:

a) The maximum exposure is defined as that level such that daily work in the environment will not result in any deviation in the normal state as well as not result in disease. Temporary changes in conditional responses are considered deviations from normal.

b) Standards are based entirely on presence or absence of biological effects without regard to the feasibility of reaching such levels in practice.

c) The values are maximum exposures rather than time-weighted averages.

d) Regardless of the value set, the optimum value and goal is zero.

e) Deviations above maximum permissible exposures *within reasonable limits* are permitted.

f) Maximum permissible exposure levels represent desirable values or ideals for which to strive.

Table 21. Radiofrequency and microwave occupational exposure limits in USSR

Frequency (MHz)	Power density (W/m²)	Electric field strength (V/m)	Magnetic field strength (A/m)	Exposure duration
0.06-1.5	-	50	5	Working day
1.5-3	-	50	-	Working day
3-30	-	20	-	Working day
30-50	-	10	0.3	Working day
50-300	-	5	0.15	Working day
300-300,000	0.25	-	-	Working day (Station.Antenna)
	1.0	-	-	2 h/day (Station.Antenna)
	10	-	-	20 min/day (Station.Antenna)
	1.0	-	-	Working day (Rotating Antenna)
	10	-	-	2 h/day (Rotating Antenna)

Table 22. USSR radiofrequency and microwave public exposure limits

Frequency (MHz)	E-field strength (V/m)	Power density (W/m²)
0.03-0.3	25	-
0.3-3	15	-
3-30	10	-
30-300	3	-
300-300,000	-	0.1

Recently, however, Sliney et al. (1985) examined the difference between the biological data that form the basis for the USSR and US standards for safety with respect to microwave radiation. According to these authors, the differences are attributed to the use of pulsed data by the Soviets as opposed to largely CW data used by the Americans.

Product emission standards

The *Radiation Control for Health and Safety Act* of 1968 (PL 90-602), administered by HEW/FDA (BRH), provides authority for controlling radiation from electronic devices in USA. The BRH microwave oven standard, effective October 1971, states that microwave ovens may not emit (leak) more than 10 W/m² at time of manufacture and 50 W/m² subsequently, for the life of the product, measured at a distance of 5 cm and under conditions specified in the standard. The Canadian standard restricts the maximum leakage to 10 W/m² at 5 cm from the oven. These standards have been adopted in Japan and most of Western Europe and endorsed by the International Electrotechnical Commission (IEC) in 1976 (Michaelson and Lin, 1987).

INTERNATIONAL ORGANIZATIONS

European Community

Since 1970 the Commission's Directorate-General for Employment, Social Affairs and Education of the European Communities has concerned itself with the influence of non-ionizing radiation on human health.

In its Sixth General Report for 1972 the European Parliament called on the Commission to establish basic standards for health protection against the danger of microwave radiation.

The first meeting of a Working Group in this field took place on March 1973. The Working Group drew up a proposal for a Commission recommendation. At the Commission's request the recommendation was converted into a proposal for a Directive and was submitted to the Council on 26 June 1980. This proposal was published in the Official Journal of the European Communities on 26 September 1980 and is entitled *Proposal for a Council Directive laying down basic standards for the health protection of the workers and the general public against the dangers of microwave radiation.* The European Parliament and the Economic Social Committee gave favourable opinions on 19 June and 25 March 1981 respectively, and the Council of Ministers instructed the Working Group on Social Affairs to prepare an opinion on this Proposal for a Directive. The structure of the Proposal for a Directive on microwave radiation was modelled on that of the Directive on ionizing radiation. The Working Group on Social Affairs had already approved several Council Directives on the protection of workers against the hazards related to chemical, physical and biological substances at the workplace and in July 1983 the Social Affairs Working Group suggested that the proposal for a Directive on microwave radiation be aligned with this model. The Commission modelled the Directive to incorporate the Council's request, with a tendency to accept that occupational exposure should not exceed a SAR of 0.4 W/kg averaged over any period of six minutes and over the whole body, or a SAR of 4 W/kg for one gramme of tissue. The discussion on which degree of conservatism had to be applied in deriving the working limits of electric and magnetic field strengths and power density from these SAR values never reached a conclusion and, currently, no Directive is expected for at least 5 years.

WHO and IRPA

Following the recommendations of the United Nations Conference on the Human Environment held in Stockholm in 1972, and in response to a number of World Health Resolutions, and the recommendation of the Governing Council of the United Nations Environment Programme, a programme on the integrated assessment of the health effects of environmental pollution was initiated by World Health Organization (WHO) in 1973. The programme, known as the WHO Environmental Criteria Programme, has been implemented with the support of the United Nations Environment Programme (UNEP) and incorporated into the International Programme in Chemical Safety. The main result of the Environmental Health Criteria Programme is a series of criteria documents.

The International Radiation Protection Association (IRPA) initiated activities concerned with Non-Ionizing Radiation by forming a Working

Group on Non-Ionizing Radiation in 1974. This Working Group later became the International Non-Ionizing Radiation Committeee (IRPA/INIRC) at the IRPA International Congress in Paris in 1977. The IRPA/INIRC reviews the literature on non-ionizing radiation and makes assessment of the health risks of human exposure to such radiation.

Two joint WHO/IRPA Task Groups on Environmental Health Criteria Documents for Extremely Low Frequency Fields (WHO/IRPA, 1984) and Radiofrequency and Microwave (WHO/IRPA, 1981) respectively, reviewed and revised existing scientific literature. They made an evaluation of the health risks of exposure to electromagnetic fields, considered rationales for the development of human exposure limits, and gave advice and recommendations for further research, so that on the basis of more definite data bases a critical revision of the existing standards can take place on an international basis, possibly reaching unique and certain safety criteria.

Various activities have been also carried out by the WHO Regional Office for Europe in the framework of a Long-Term Programme on Non-ionizing Radiation Protection; one of the important results achieved has been the publication of a manual on health aspects of exposure to non-ionizing radiation also covering legislation and regulations in the radiofrequency and microwave range (WHO, 1982).

WHO and IRPA have recently proposed new criteria and standards for limiting the exposure of workers and the general population to radiofrequency and microwave radiations.

WHO suggests (WHO/IRPA, 1981) that exposures to power densities in the range 1-10 W/m^2 are acceptable for occupational exposure throughout a complete working day and that higher exposures might be reasonable for some frequency ranges and occasional exposure. For the general public it is suggested that lower, but unspecified, exposure levels are appropriate and that the levels should be as low as reasonably achievable, according to the ALARA principle.

The IRPA recommendations (IRPA, 1988) are more specific and have the purpose of providing guidance on limits of exposure to electromagnetic radiation and fields in the frequency range from 100 kHz to 300 GHz. In the IRPA guidelines, the basic limits of exposure formulated for the frequency region of 10 MHz and above are expressed by the quantity specific absorption rate, SAR, i.e. the power absorbed per unit mass. In the frequency region below 10 MHz, basic limits are expressed in terms of the effective electric and magnetic field strengths; in the near field or in multipath fields, both the electric and magnetic field strengths must be measured.

Occupational exposure to RF radiation at frequencies below and up to 10 MHz should not exceed the levels of unperturbed rms electric and magnetic field strengths given in Table 23, when the squares of the electric and magnetic field strengths are averaged over any 6-minute period during the working day, provided that the body-to-ground current does not exceed 200 mA, and that any hazard of RF burns is eliminated according to the recommendations stated below. A conservative approach consists in limiting pulsed electric and magnetic field strengths as averaged over the pulse width to 32 times the values given in Table 23.

Occupational exposure to frequencies above 10 MHz should not exceed a SAR of 0.4 W/kg when averaged over any 6-minute period and over the whole body, provided that in the extremities (hands, wrists, feet and ankles) 2 W per 0.1 kg shall not be exceeded and that 1 W per 0.1 kg shall not be exceeded in any other part of the body.

Table 23. IRPA occupational exposure limits to radiofrequency electromagnetic fields

Frequency range (MHz)	Unperturbed rms electric field strength (V/m)	Unperturbed rms magnetic field strength (A/m)	Equivalent plane wave power density (W/m²)	(mW/cm²)
0.1-1	614	1.6/f	-	-
>1-10	614/f	1.6/f	-	-
>10-400	61	0.16	10	1
>400-2,000	$3f^{0.5}$	$0.008f^{0.5}$	f/40	f/400
>2,000-300,000	137	0.36	50	5

f = frequency in MHz
Note: Hazards of RF burns should be eliminated by limiting currents from contact with metal objects. In most situations this may be achieved by reducing the E values from 614 to 194 V/m in the range from 0.1 to 1 MHz and from 614/f to $194/f^{0.5}$ in the range from >1 to 10 MHz.

The limits of occupational exposure given in Table 23 for the frequencies between 10 MHz and 300 GHz are the IRPA working limits derived from the SAR value of 0.4 W/kg. They represent a practical approximation of the incident plane wave power density needed to produce the whole body average specific absorption rate of 0.4 W/kg. These limits apply to whole body exposure from either continuous or modulated electromagnetic fields from one or more sources, averaged over any 6-minute period during the working day (8 hours per 24 hours).

Although very little information is presently available on the relation of biological effects with peak values of pulsed fields, it is suggested that the equivalent plane wave power density as averaged over the pulse width not exceed 1000 times the power density limits in Table 23 for the frequency concerned, provided that the limits of occupational exposure, averaged over any 6-minute period, are not exceeded.

For the case of the near field, where a complex phase relationship between the magnetic and electric field components exists, it is possible that exposure may be predominantly from either one of these components, and in extreme cases from the magnetic or electric field alone.

The limits for magnetic and electric field strengths indicated in Table 23 for frequencies above 10 MHz may be exceeded for the case of near field exposure, provided that:

$$2.2 \times 10^{-3} E^2 + 62.83 H^2 < P$$

where E is the electric field strength (V/m), H is the magnetic field strength (A/m), and P is the equivalent plane wave power density limit (W/m²) from Table 23, and the SAR limits of occupational exposure, averaged over any 6-minute period, are not exceeded. The above formula may be applied in practical situations to the case of near field exposures in the frequency range from 10 MHz to 30 MHz, in rare instances up to 100 MHz.

Table 24. IRPA general public exposure limits to radiofrequency electromagnetic fields

Frequency range (MHz)	Unperturbed rms electric field strength (V/m)	Unperturbed rms magnetic field strength (A/m)	Equivalent plane wave power density (W/m^2)	(mW/cm^2)
0.1-1	87	$0.23/f^{0.5}$	-	-
>1-10	$87/f^{0.5}$	$0.23/f^{0.5}$	-	-
>10-400	27.5	0.073	2	0.2
>400-2,000	$1.375f^{0.5}$	$0.0037f^{0.5}$	$f/200$	$f/2,000$
>2,000-300,000	61	0.16	10	1

f = frequency in MHz.

Exposure of the general public to RF radiation at frequencies below and up to 10 MHz should not exceed the levels of unperturbed rms electric and magnetic field strengths given in Table 24, provided that any hazard of RF burns is eliminated according to the recommendations stated below.

Radiofrequency radiation exposure of the general public at frequencies above 10 MHz should not exceed a SAR of 0.08 W/kg when averaged over any 6 minutes and over the whole body. The limits of RF exposure to the general public given in Table 24 for the frequencies between 10 MHz and 300 GHz are derived from the SAR value of 0.08 W/kg. They represent a practical approximation of the incident plane wave power density needed to produce the whole body average specific absorption rate of 0.08 W/kg. These limits apply to whole body exposure from either continuous or modulated electromagnetic fields from one or more sources, averaged over any 6-minute period during the 24-hour day.

Although very little information is presently available on the relation of biological effects with peak values of pulsed fields, it is suggested that the equivalent plane wave power density as averaged over the pulse width not exceed 1000 times the power density limits or the field strengths not exceed 32 times the limits given in Table 24 for the frequency concerned, provided that the limits for public exposure, averaged over any 6-minute period, are not exceeded, and hazards of RF burns are eliminated.

Radiofrequency shocks and burns can result from touching ungrounded metal objects that have been charged up by the field or from contact of a charged up body with a grounded metal object. If the current at the point of contact exceeds 50 mA, there is a risk of burns.

In view of the lack of a complete understanding of the biophysics of living systems, and the controversies concerning the non-thermal electromagnetic field interactions, no predictive theory based on non-thermal considerations is possible. Empirical data must serve for establishing exposure limits. According to Schwan (1982b), non-thermal effects may play a significant role at frequencies below 0.1 MHz. No adequate data on biological effects at these frequencies exist, except for the extremely low frequency range (WHO/IRPA, 1984). Therefore it seemed to be prudent to limit the lower frequency end of the guidelines to 0.1 MHz. If compelling practical reasons (wide application at a particular frequency) exist, a limit for a narrow range or a particular device (emission standard) operating at a frequency below 100 kHz can be set on the basis of empirical studies.

Non-thermal effects are expected (Illinger, 1982) to be frequency-dependent and may exhibit complex dose effect relationships, including intensity windows both at low and high (above 30 GHz) frequencies (Grundler et al., 1982; Schwan, 1982b). Threshold data on adverse health effects applicable to humans over the whole frequency range and to all possible modulations do not exist. Most of the present biological data exist for frequencies in the range from 900 MHz to 10 GHz. Thus, where data are lacking, assumptions on possible adverse health effects must be made. These are based on our present knowledge of biophysics of electromagnetic radiation absorption and on analytical or experimental models, as well as on limited epidemiological data. Extrapolation of limits in the range 10 MHz to 300 GHz is based on the frequency dependence of energy absorption in human beings. This may account for thermal effects but it does not allow one to predict non-thermal effects.

CONCLUSIONS

When one compares all current radiofrequency/microwave standards internationally, a paradox contained in all these standards activities is quite apparent. Presumably, each of these national groups had available to it the same library of reference data, perhaps supplemented to a small degree by some preparatory investigational and research results. And yet the permissible limits for occupational exposures differ by factors of 20 to 100, and those for the general public by factors of 20.

Among the factors that contribute to the extreme differences among standards intended for the control of human exposure to RF/MW radiation, five are of major importance: (a) the particular physical and biological effects data selected as the basis for standards development, (b) the interpretation of these data into terms significant to humans and their environment, (c) the different purposes to be served by the standards, (d) the compromises made between levels of risk and degrees of conservatism, and (e) the influence of preceding standards activities in each particular nation and in neighboring areas having allied socio-political outlooks.

Governments and industry alike are being compelled to adopt RF/MW safety standards, in spite of the fact that the scientific data base is neither adequate nor complete, nor can much of it be referred usefully to the human situation. The quality and comprehensiveness of the scientific data base determine the confidence with which standards can be derived and defended. To make its data useful to standards-setters, the scientific community needs to agree on descriptive methods of accounting for the wavelength and frequency dependent properties of exposed subjects that determine the patterns of internal RF/MW energy deposition. To make results from related studies intercomparable, common techniques and terminology for internal dosimetry need further development, agreement, and use.

The acceptance of whole-body averaged SAR in the development of the new frequency-dependent safety standards has been a significant improvement. Despite this fact, whole-body SAR seems not an adequate basis for safety guidelines at frequencies greater than 20 GHz and less than 3 MHz. At frequencies greater than 20 GHz, microwave energy deposition in biologic tissue is very superficial, and some form of localized SAR would serve as a better safety guideline than a whole-body averaged SAR. Also, whole-body average SAR is not an adequate basis for the safety guidelines below about 3 MHz since absorption in biologic tissue decreases as a function of frequency squared below the resonant frequency, becoming energy absorption small in this frequency range. Valid questions remain, however, concerning the potential for shocks and burns, as treated by IRPA, ACGIH, NRPB, and Australian standards.

In conclusion, different scientific approaches have produced different philosophies of protection guidelines and thus different exposure limits. It is however apparent that, in the light of the continuous advancement of scientific results, the differences are decreasing and the revisions of existing standards or the setting of new ones reflect at least the tendency to merge to a common area.

It is important, however, to maintain an adequate research effort so that on the basis of more definite data bases a periodical critical revision of the existing standards can take place on an international basis, possibly reaching unique and certain safety criteria.

The international cooperation in the development of compatible standards should be encouraged, because the lack of international agreement on the protection standards to be used for non-ionizing radiation (Repacholi, 1983; Michaelson, 1983; Duchene and Komarov, 1984) constitutes a major drawback for the development of safety regulations in those countries where they do not yet exist. Therefore, it must be hoped that the efforts outlined above to achieve international cooperation in the field of non-ionizing radiation together with progress in our knowledge on the biological effects will allow protection against non ionizing electromagnetic fields to develop in the same climate of international agreement as that which has been reached for ionizing radiation.

REFERENCES

ACGIH, 1988, "Threshold Limit Values and Biological Exposure Indices for 1988-89," American Conference of Governmental Industrial Hygienists, Cincinnati.
Allen, S. G., and Harlen, F., 1983, "Sources of Exposure to Radiofrequency and Microwave Radiation in the UK," Report NRPB-R144, National Radiological Protection Board, Chilton, Didcot.
ANSI, 1982, "American National Standard-Safety Levels with Respect to Human Exposure to Radio Frequency Electromagnetics Fields, 300 kHz to 100 GHz," Report ANSI C95.1-1982, American National Standards Institute, New York.
Baranski, S., and Czerski, P., 1976, "Biological Effects of Microwaves," Dowden, Hutchinson & Ross Editors, Stroudsburg.
Bernhardt, J. H., 1985, Evaluation of human exposure to low frequency fields, in: "The Impact of Proposed Radio-frequency Radiation Standards on Military Operations," AGARD lecture series No 138, NATO AGARD, Neuilly sur Seine.
Counts, D. L., 1982, "Non Ionizing Radiation Program at Lawrence Livermore National Laboratory," Proc. of III Int. Congress of the Society for Radiological Protection, Inverness.
Czerski, P., 1985, Radiofrequency radiation exposure limits in Eastern Europe, J. Microwave Power, 20:233.
Czerski, P., and Piotrowski, M., 1972, Proposal for specifications of allowable levels of microwave radiation, Medycyna Lotnicza, 39:127 (In Polish).
DIN VDE, 1986, "Hazards by Electromagnetic Fields: Protection of Persons in the Frequency Range from 0 Hz to 3,000 GHz," Report DIN VDE 0848, part 2 (In German).

Duchene, A., and Komarov, E., 1984, "International Programmes and Management of Non-ionizing Radiation Protection," Proc. of IRPA 6th International Congress, Berlin.

Durney, C. H., Massoudi, H., and Iskander, M. F., 1986, "Radiofrequency Radiation Dosimetry Handbook," 4th ed., Report SAM-TR-85-73, USAF School of Aerospace Medicine, Brooks Air Force Base, Texas.

Elder, J. A., and Cahill, D. F., eds, 1984, "Biological Effects of Radiofrequency Radiation," US Environmental Protection Agency, Research Triangle Park.

Gandhi, O. P., 1986, Recent advances in the dosimetry of radiofrequency and microwave radiation, in: "Physics in Environmental and Biomedical Research," S. Onori, and E. Tabet, eds., World Scientific Publishing Co., Singapore.

Gandhi, O. P., Haymann, M. J., and D'Andrea, J. A., 1979, Part-body and multi-body effects on absorption of radiofrequency electromagnetic energy by animals and by models of man, Radio Science, 14:15.

Glaser, Z. R., 1980, "Basis for the NIOSH radiofrequency and microwave radiation criteria document," Proc. of an ACGIH Topical Symposium on Non-Ionizing Radiation.

Grandolfo, M., 1985, What data bases for electromagnetic fields exposure standards? in : "Interactions between Electromagnetic Fields and Cells," A. Chiabrera, C. Nicolini, and H. P. Schwan, eds, Plenum Press, New York.

Grandolfo, M., 1986, Occupational exposure limits for radiofrequency and microwave radiation, Appl. Ind. Hyg., 1:75.

Grandolfo, M., Michaelson, S. M., and Rindi, A., eds, 1983, "Biological Effects and Dosimetry of Nonionizing Radiation: Radiofrequency and Microwave Energies," Plenum Press, New York.

Gronhaug, K. L., and Busmundrud, O., 1982, "Antenna Effect of the Human Body to EMP," Report FFIF/453/153, Norwegian Defence Research Establishment, Kjeller (In Norwegian).

Grundler, W., Keilmann, F., Putterlik, V., and Strube, D., 1982 Resonant-like dependence of yeast growth rate on microwave frequencies, Brit. J. Cancer., 45(supp.V): 206.

Hankin, N. H., 1986, "The Radiofrequency Radiation Environment: Environmental Exposure Levels and RF Radiation Emitting Sources," EPA 520/1-85-014, Environmental Protection Agency, Las Vegas.

Hill, D. A., 1984, The effect of frequency and grounding on whole-body absorption of humans in E-polarized radiofrequency fields, Bioelectromagnetics, 5:131.

HSE, 1986, "Safety in the Use of Radiofrequency Dielectric Heating Equipment," Guidance Note PM51, Health and Safety Executive, Plant and Machinery, UK.

HWC, 1979, "Recommended Safety Procedures for Installation and Use of Radiofrequency and Microwave Devices in the Frequency Range 10 MHz-300 GHz," Publication 79-EHO-30, Health and Welfare Canada, Ottawa.

Illinger, K. H., ed., 1982, "Biological Effects of Nonionizing Radiation," ACS Symposium Series No.157, American Chemical Society, Washington.

INRS, 1978, "Le Rayonnement Electromagnetique "Radiofrequences" Application et Risques", ND 1127-92-78 (39-14) CDU 621.37: Cahiers de notes documentaires, No 92, 3e Trimestre 1978.

INRS, 1982, "Risques Lies aux Rayonnements Electromagnetiques Non Ionisants," Cahiers de notes documentaires, No 197, 2e Trimestre 1982.

INRS, 1983, "Valeurs Limites d'Exposition aux Agents Physiques", Cahiers de notes documentaires, No 110, 1er Trimestre 1983.

INRS, 1985, "Protection contre les Rayonnements Electromagnetiques Non Ionisants," ND 1552-121-85 CDU 538.56. Cahiers de notes documentaires No 121, 4e Trimestre 1985.

IRPA, 1984, Interim guidelines on limits of exposure to radiofrequency electromagnetic fields in the frequency range from 100 kHz to 300 GHz, Health Physics, 46:975.

IRPA, 1988, Guidelines on limits of exposure to radiofrequency electromagnetic fields in the frequency range from 100 kHz to 300 GHz, Health Physics, 54:115.

Jokela, K., 1988, Private communication.

Magnuson, H. T., 1964, Industrial toxicology in the Soviet Union - theoretical and applied, Am. Ind. Hyg. Assoc. J., 25:185.

Mahra, K., 1970, Maximum admissible levels of HF and UHF electromagnetic radiation at work places in Czechoslovakia, in "Biological Effects and Health Implications of Microwave Radiation," S. F. Cleary, ed., Publication BRH/DBE 70-2, US Dept. of Health, Education, and Welfare, Rockville.

Michaelson, S. M., 1983, Microwave/radiofrequency protection guide and standards, in: "Biological Effects and Dosimetry of Nonionizing Radiation: Radiofrequency and Microwave energies," M. Grandolfo, S. Michaelson, and A. Rindi., eds., Plenum Press, New York.

Michaelson, S. M., and Lin, J. C., 1987, "Biological Effects and Health Implications of Radiofrequency Radiation," Plenum Press, New York.

Mitchell, J. C., 1985, Development and application of new radiofrequency radiation safety standards, in: "The Impact of Proposed Radio-frequency Radiation Standards on Military Operations," AGARD lecture series No 138, NATO AGARD, Neuilly sur Seine.

Musil, J., 1983, "The Second Draft of New Czechoslovakian Standard," Abstracts 5th Annual Session of the Bioelectromagnetics Society, Boulder.

NRPB, 1982, "Proposal for the Health Protection of Workers and Members of the Public against the Dangers of Extra Low Frequency, Radiofrequency and Microwave Radiations. A Consultative Document," ISBN 085951 185 5, National Radiological Protection Board, Chilton, Didcot.

NRPB, 1986, "Advice on the Protection of Workers and Members of the Public from the Possible Hazards of Electric and Magnetic Fields with Frequencies below 300 GHz. A Consultative Document," ISBN 085951 267 3, National Radiological Protection Board, Chilton, Didcot.

ONORM, 1986, "Microwave and Radiofrequency Electromagnetic Fields; Definitions, Limits of Exposure, Measurements," Onorm S1120, Osterreichisches Normungsinstitute, Vienna (In German).

Petrov, I. R., and Subbota, A. G., 1972, Concluding remarks, in: "Influence of Microwave Radiation on the Organism of Man and Animals," I. R. Petrov, ed., NASA Translation TT-F-108.

Polk, C., and Postow, E., eds., 1986, "CRC Handbook of Biological Effects of Electromagnetic Fields," CRC Press, Boca Raton.

Repacholi, M. H., 1978, Proposed exposure limits for microwave and radiofrequency radiation in Canada, J. Microwave Power, 13:199.

Repacholi, M. H., 1983, Development of standards - Assessment of health hazards and other factors, in: "Biological Effects and Dosimetry of Nonionizing Radiation: Radiofrequency and Microwave Energies," M. Grandolfo, S. M. Michaelson, and A. Rindi, eds., Plenum Press, New York.

Rozzell, T. C., 1985, West Germany EMF exposure standard, BEMS Newsletter, 55.

SAA, 1985, "Maximum Exposure Levels - Radiofrequency Radiation - 300 kHz to 300 GHz," Australian Standard 2772, The Standards Association of Australia, Sydney.

Saxebol, G., 1982, "Administrative Normer for Radiofreqvent Straling for Yrkeseksponerte?," SIS Rapport 1982:2, State Institute of Radiation Hygiene, Oslo (In Norwegian).

Schwan, H. P., 1982a, Microwave and RF hazard standard considerations, J. Microwave Power, 17:1.

Schwan, H. P., 1982b, Nonthermal cellular effects of electromagnetic fields: AC-field induced ponderomotoric forces, <u>Brit. J. Cancer</u>, 45(supp.V):220.

Sliney, D. H., Wolbarsht, M. L., and Muc, A. M., 1985, Differing radio frequency standards in the microwave region; implications for future research, <u>Health Physics</u>, 49:677.

Stuchly, M. A., 1987, Proposed revision of the Canadian recommendations on radiofrequency-exposure protection, <u>Health Physics</u>, 53:649.

USSR, 1976, USSR Standard for Occupational Exposure, Amendment N.1 of January 1st, 1982, to State Standard 12.1.006-76

WHO/IRPA, 1981, "Radiofrequency and microwaves," Environmental Health Criteria 16, WHO, Geneva.

WHO/IRPA, 1984, "Extremely Low Frequency (ELF) Fields," Environmental Health Criteria 35, WHO, Geneva.

WHO/Regional Office for Europe, 1982, "Nonionizing Radiation Protection," WHO Regional Publications, European series No. 10, Copenhagen (1982).

THE RATIONALE FOR THE EASTERN EUROPEAN RADIOFREQUENCY

AND MICROWAVE PROTECTION GUIDES

Stanislaw Szmigielski, and Tadeusz Obara

Department of Biological Effects of Non Ionizing Radiation
Center for Radiobiology and Radiation Safety
128 Szaserow, 00-909 Warsaw, Poland

INTRODUCTION

The safety standards for environmental and occupational exposure to electromagnetic fields (EMFs) being accepted and operating in Eastern European countries, members of the Council for Mutual Economic Assistance (Comecon), and in the USSR follow since their establishment a different philosophy, as compared to the standards and rationales introduced in the USA and Western European countries.

Historically, in the early 1950s, the safety levels for uncontrolled occupational exposure of personnel to microwaves (MWs) were set in the USSR at 10 $\mu W/cm^2$ (0.1 mW/m^2) and this value was soon accepted in other Eastern European countries, influencing the radiofrequency (RF) protection guides in these countries up to very recently. The basis of this standard was claimed by earlier Soviet authors (Gordon, 1974) on appearance of "pathologic" changes in experimental animals exposed to 2,450 MHz MWs at power densities of 100-1,000 $\mu W/cm^2$ and the introduction of the "safety coefficient" of 10-100.

From that time opinions of Soviet authors concerning biological effects and health hazards of RF radiation have changed considerably (Akoev, 1986; Akoev et al., 1986; Shandala, 1986) but the "old" standard of 10 $\mu W/cm^2$ for uncontrolled exposure to MWs still influences the present protection guides and rationales. The present safety standards and protection guides in the USSR and Eastern European countries are presented and summarized in the present book (see Grandolfo and Mild's chapter) and there is no sense to repeat the values here. Thus, the aim of the present review is to describe and critically discuss the rationales and tendencies for amendments of EMF safety rules in Poland and in other Eastern European countries.

Despite numerous experimental investigations and epidemiological studies, it is still not possible to prove the existence and character of any specific molecular, cellular or system-related damage that may be evoked by long-term exposure to EMFs. Excluding the well defined thermal effects and reactions to local or whole body hyperthermia, it became evident from the beginning of bioelectromagnetic studies that biological effects to single or long-term exposures in nonthermal EMFs are inconsistent, transient and difficult to confirm and interpret. For biological effects related to EMFs it is not possible to determine the primary target organ or system that may be considered as "biological dosimeter" in terms of health hazards.

On the other side, under certain well controlled conditions of exposure in experimental EMFs, a variety of behavioural, neurological, endocrine, hemato-immunological as well as reproductive abnormalities were demonstrated (for review, see Davidov et al., 1984; Elder and Cahill, 1984). However, it is still not possible to establish valid thresholds of field power density or energy absorption for the above phenomena. Medical examinations of personnel exposed occupationally to EMFs and epidemiological analyses of large groups of people working in this type of environment also did not show conclusive health hazards related to EMFs, although a variety of non-specific symptoms, including liability of vegetative nervous system and increased frequency of neuroses, have been reported (Baranski and Czerski, 1976; Elder and Cahill, 1984; Minin, 1974).

This situation has led to the formulation of a hypothesis recognising EMFs as a non-specific chronic stressor with the partly documented assumption that non-ionizing radiation can be detected by the peripheral and/or central nervous system. The nervous system, both directly or through its physiological control-regulatory connections with the immune and endocrine systems, may evoke adaptive and/or compensatory reactions that under certain conditions (or in genetically determined subjects) result in the development of a disease. Thus, chronic exposure to EMFs may be considered as a risk factor for disease. The hypothesis of EMFs as a non-specific stressor, popular among Soviet and Eastern European authors, was recently summarized by Marino (1988).

Lack of generally accepted and repeatable biological effects and health hazards due to long-term exposures to low-level EMFs, considerably impedes establishing the satisfactory safety levels and maximum permissible intensities of EMFs that may be considered safe for occupational exposure. The present situation is additionally complicated by the fact that virtually nothing is known for sure about possible long-term and delayed effects of exposure at different levels of EMFs intensity (e.g. life span, carcinogenesis, influence of future generations) as well as about biological effects of complex EMFs and combined action of EMFs and other harmful environmental or occupational risk factors. For example, the experimental data indicating that MW radiation (not being carcinogenic per se), may enhance a potence of certain carcinogens, as well as the recently reported epidemiological data on increased risk of neoplasms in humans exposed to EMFs (for review, see Szmigielski and Gil's chapter in this book), should be taken in consideration as possible clues influencing future changes in safety levels of EMFs.

Taking into account all the above, it is not surprising that bioelectromagnetic experts in the USSR and in Poland adviced in early 1960s that permissible levels of EMF intensity for public and occupational exposure should be set as low as it was economically acceptable and under these conditions of exposure medical and epidemiological observations should be performed on sufficiently large populations of human beings.

The results of these observations could demonstrate possible health disorders and any tendency toward development of systemic diseases that may be related to exposure to EMFs. This in turn could allow to assess health risks connected with long-term exposure to EMFs, and modify accordingly the safety levels. This is the basic rationale for EMFs safety standards in Eastern European countries, and when they were introduced originally in the 1960s many of the experts were well aware that the operating safety levels were (or not!) unnecessarily low and could or would be elevated in the future.

RATIONALES FOR ESTABLISHING EMF SAFETY RULES

Data being acceptable for considerations concerning safety levels of EMFs are derived from:

- theoretical considerations of power absorption and possible interactions with living materials;
- experimental investigations of cells or whole organisms exposed to artificial EMFs under a variety of conditions and combinations;
- periodical medical examinations of personnel exposed occupationally to EMFs with a search for relations existing between the observed effects and the intensity and period of exposure;
- epidemiological analysis of health risk assessment in large groups of human beings either exposed occupationally or living in areas with environmental intensities of EMFs above the average.

Unfortunately, none of these sources of information provide data that may be directly used as sure indicators of safety levels and thus, in any case, the decision at which level of EMF intensity should the bar be settled is taken to some extent arbitrarily.

Discussing rationales for EMF safety levels a possibility of certain interactions connected with pulse modulation of the carrier wave should be also considered. Early Soviet and Eastern European literature (Baranski and Czerski, 1976; Minin, 1974) report opinions that pulse modulated MW radiation exerts stronger neurological, behavioural and hemato-immunological effects, compared to continuous wave of the same frequency. Actually, there are no convincing evidences that millisecond pulses of MWs at mean power densities not leading to detectable thermal effects may influence function of living organisms to a higher degree than continuous wave. However, certain cellular disturbances attributed to specific interactions of RF/MWs modulated sinusoidally at the 1-100 Hz frequencies have been recently postulated (Adey, 1981; Byus et al., 1984, 1986). These authors have gained numerous evidences that the amplitude modulation characteristics appear to be a prime determinant of the nature of interactions at the cellular level with "windowing" of many of the observed interactions with both the frequency and amplitude of pulses. These and other not fully understandable and unresolved events occurring in cells exposed to low level RF/MW modulated fields underline the complexity of interactions and suggest caution in establishing the safety levels.

There exists at present a general agreement among Polish experts that in case of MW radiation, either continuous or pulse-modulated, no significant shifts in function of the organism of experimental animals occur at power densities below 1 mW/cm^2 (10 W/m^2), even in mice exposed to the resonant frequency of 2,450 MHz. In early Soviet and Eastern European literature (summarized by Baranski and Czerski, 1976; Minin, 1974) there are numerous reports on a variety of behavioural, neurological, biochemical and hemato-immunological effects occurring in animals exposed for a long time to extremely low (1 µW/cm^2) MW fields. However, in a recent Soviet monograph (Davidov et al., 1984) alterations occurring after exposure of mice and rats to very low MW fields are no longer presented and discussed. Instead, the authors claim that inconsistent and transient immunological alterations may be observed in animals exposed for a longer time to power densities exceeding 0.5

mW/cm^2, while clearly demonstrable immunological effects are observed only at thermogenic power density.

Another important thing to keep in mind in defining safety criteria originating from experimental investigations of animals exposed to EMFs, is the lack of linear correlations between the observed biological effects and intensity of EMFs, at least in the RF/MW range. For example, the reaction of the immune system to the exposure to RF/MW fields (for the most recent review, see Szmigielski et al., 1988) appears to be biphasic with a phase of immunostimulation, followed by transient suppression of certain immune reactions.

Soviet authors (Savin et al., 1983; Akoev, 1986) in the recent guidelines for the revision of RF/MW safety criteria have also stressed the non-linear correlation between observed biological effects and field power density. These authors claim that non-linear dependences occur at power densities of 4-10 mW/cm^2 with closer linearity below 4 mW/cm^2.

Medical examinations of personnel exposed occupationally to EMFs, at least those performed till now in Poland, are still not fully conclusive in terms of safety criteria, although all people who begin to work in EMF environments undergo general, ophtalmological and neurological medical examinations that are later (during employment period) repeated every 1-3 years. The syndrome of "microwave disease" reported in earlier Soviet and Polish literature (Baranski and Czerski, 1976) is no longer diagnosed in Poland, although recent medical statistics in this country indicate the occurrence of more frequent vegetative neuroses and a variety of non-specific symptoms (liability of blood pressure, fatigue, headaches, etc.) in personnel occupationally exposed to MW/RF radiation. However, these symptoms do not show a close relation to period of employment or intensity of exposure.

Considering the health hazards related to occupational exposure to EMFs, Polish medical experts stress two important facts from the point of view of safety criteria:

1. Presence of single subjects being hypersensitive to very low EMFs in a wide range of frequencies;
2. More frequent occurrence of non-specific symptoms during the first one-two years of employment in EMF environments with disappearance of the symptoms during further years of employment.

Both these observations cannot be still convincingly proved statistically, but there exists a general agreement that both should be taken into account in the selection of personnel for employment in EMF environments. Subjects diagnosed as hypersensitive to EMFs suffer from a variety of non-specific but severe symptoms (fatigue, migraines, speech problems, depression, rarely convulsions) when exposed to low (microwatt-milliwatt) MW/RF fields.

Some of these subjects suffer also from similar symptoms when they find themselves in the vicinity of power lines, high power TV/Radio transmitters or even near VDTs. In our findings the hypersensitivity to EMFs, still being a nonunderstandable phenomenon, occurs rarely (in a few cases per 10,000 human beings) while severe hypersensitivity is very exceptional (we observed only two cases). Recently, similar observations on hypersensitivity to EMFs were reported by Choy et al. (1987).

In summary, there still exist many unresolved and/or uncertain issues in the assessment of biological effects and possible health hazards related both to environmental and occupational exposure to EMFs. The major problems are the following:

- thresholds for specific biological effects (non linear relation of bioeffects);
- biological effects of complex and modulated fields;
- interaction of EM fields with other risk factors;
- specific effects of EM fields versus non specific stress reaction;
- long-term and/or delayed effects of EM fields (e.g. carcinogenesis);
- influence of EM fields on development and course of other diseases and interaction with applied therapy (e.g. drugs).

The present knowledge in this field, although based on relatively broad experimental and medical data, does not provide sufficient background for establishing satisfactory and generally acceptable safety levels, at least until the unresolved above listed problems will be better defined. Therefore, all safety levels for EMFs are established arbitrarily taking into account sufficient "safety coefficients" to avoid the appearance of unsuspected and/or still unknown health effects. Certainly, the arbitrarily established safety levels should be amended with the enhancement of knowledge in bioelectromagnetics.

In practice, there are two main problems in establishing safety criteria for EMFs that in our opinion should be considered separately and with distinct philosophies:

- environmental (public) uncontrolled exposure;
- occupational exposure.

Considering the safety standards proposed or operating in various countries we have the feeling that these two kinds of exposure are not sufficiently differentiated and in some cases there is a tendency to establish one level for both.

ENVIRONMENTAL SAFETY CRITERIA

For a very long time exposure of living organisms to non-ionizing radiation was only due to natural EMFs. With the appearance of artificial EMFs, the environmental exposure has increased by a few orders of magnitude, and due to the rapid growth of both number and power of devices generating EMFs directly in the atmosphere as well as to reflections from buildings, hills, metallic constructions, etc., there are at present places where the intensity of EMFs at various frequencies has reached a value of from a few to several $\mu W/cm^2$. The existing safety levels for uncontrolled environmental exposure to MWs are still a matter of controversy, as the environment is characterized by very complex fields, difficult to measure, predict and interpret.

Environmental exposure covers in practice all frequencies of EMFs (Table 1) including ELF, LF, RF and MWs and a variety of modulations, although in most cases a certain and measurable frequency predominates, depending upon vicinity of EMF sources. On the contrary, there is a tendency toward establishing safety levels for single ranges of EMFs, at least for ELF, RF and MWs (Table 1). As the present knowledge on biological effects of complex EMFs and on interaction of weak EMFs with other environmental risk factors (e.g. chemical pollutants, carcinogens) is very scarce and fragmentary, the only logical, although tentative, solution is to apply a "safety coefficient" of about 10-100.

Table 1. Spectrum of electromagnetic fields from the point of view of safety levels in USSR

	Frequency ranges	EM radiation
1	Below 30 Hz	Extremely Low Frequency (ELF)
2	30-300 Hz	Extremely low Frequency (ELF)
3	300 Hz-3 kHz	Low Frequency (LF)
4	3-30 kHz	Low Frequency (LF)
5	30-300 kHz	Low Frequency (LF)
6	300 kHz-3 MHz	Radiofrequencies (RF)
7	3-30 MHz	Radiofrequencies (RF)
8	30-300 MHz	Radiofrequencies (RF)
9	300 MHz-3 GHz	Microwaves (MW)
10	3-30 GHz	Microwaves (MW)
11	30-300 GHz	Microwaves (MW)

This solution is based on what may be concluded from known biological effects with an unproven hope that this will protect whole mankind, including high risk groups (e.g., children, pregnant women, etc.) against unsuspected and/or yet unknown long-term effects. In other words, the environmental safety levels for EMFs are established at "unwarranted" low levels, until more will be known on interactions of EMFs with other risk factors and on bioeffects of complex EMFs.

Certainly, the best safety level would be "zero", limiting environmental exposure to natural EMFs, but this is no longer possible in any place in the world.

Thus, a logical compromise is needed, taking into account the economic costs of the elaborated safety levels. For example, lowering the RF/MW environmental safety levels to 1 $\mu W/cm^2$, which would close large areas for building, housing, agriculture, etc., in the vicinity of radar stations, air force bases and TV/radio antennas, would be economically groundless. On the other side, establishing the RF/MW environmental safety levels at 1 mW/cm^2, which would solve most of the above problems, would be considered too high in view of possible health hazards. So, the environmental RF/MW safety levels should be arbitrarily set somewhere in the range of 1-1,000 $\mu W/cm^2$, taking into account the complexity and variability of pollution by EMFs.

Furthermore there is a need to establish safety levels for at least a few frequencies (Table 1) that evoke bioeffects at different levels of intensity. For frequencies below RFs (below 0.1-0.3 MHz) separate levels for E and H field strenghts should be considered. The last remarks are more important for occupational safety standards, but to some extent they should be taken into account also for environmental exposures.

In the environmental EMF exposure, far field (plane wave) conditions apply and thus it is foreseen for the future to establish only one integrated safety level for the whole spectrum of EMFs (except of ELFs) in the frequency range 1 kHz-300 GHz, with the only measurement of power density in the whole range.

The present tendencies in Eastern European countries are however far from this concept. For example, the recent USSR and Eastern European countries proposal for EMF environmental safety levels, sets the levels for 6 subranges (Table 1) covering only the RF/MW frequencies. In our opinion this is the echo of occupational safety standards, but the authors of this proposal did not take into account the complexity of EMFs in the environment with respect to those in the workplace.

Table 2 lists factors important for safety regulations in exposure to EMFs, stressing the differences and similarities between environmental and occupational exposures. Some of the factors differ (e.g. uncontrolled exposure of high risk groups, complex EMF in environmental exposure versus controlled exposure of preselected personnel, defined and easily measurable EMFs in occupational exposure), others are more important for environmental exposure (long-term and delayed effects, interaction with other pollutants) or predominate in occupational exposure (adaptation and transient non-specific symptoms that disappear after 1-2 years of employment) and still other are equally important for both kinds of exposures (hypersensitivity to EMFs, assessment of health risk related to EMFs). The different conditions and factors occurring in environmental and occupational exposures to EMFs strongly indicate a need for different safety levels and protection guides for the whole population and for the preselected group of personnel working in EMF environments.

Table 2. Differences and similarities between environmental and occupational exposure to EMFs.

ENVIRONMENTAL EXPOSURE	OCCUPATIONAL EXPOSURE
Uncontrolled exposure, including high risk groups	Controlled exposure of preselected personnel
Complex EM fields, difficult for prediction and measurements	Defined and measurable EM fields
Periodic epidemiological observation of population	Periodic medical examination and selection
Adaptation and transient non-specific symptoms	
Long-term and delayed effects including carcinogenesis	
Interaction with other environmental risk factors	
HYPERSENSITIVITY TO EM FIELDS ASSESSMENT OF RISK	
Need for uniform safety level covering the broadest possible range of EM spectrum	Need for safety levels and regulations covering the specified conditions of working place

OCCUPATIONAL SAFETY CRITERIA

In the case of occupational exposure, the situation appears to be different from that discussed for environmental EMF pollution (Table 2). The main differences can be summarized in the following points.

1. There is a possibility for reasonably precise characterization of EMFs occurring at the place of occupational exposure.

2. EMFs occurring in occupational exposure are normally much less complex than those in the environment and in most cases there is a very strong predominance of one frequency, although with a variety of modulations and a possibility of interferences.

3. It is possible, reasonably easy, and economically convenient to lower occupational exposure levels by means of ergonomy of the working place, screens, and individual protecting devices (suits, glasses, etc.).

4. The period of occupational exposure can be controlled and limited in time.

5. High risk groups of people (children, pregnant women, senior citizens, subjects with health disorders) are excluded from occupational exposure.

6. With proper organization of the industrial/military medical services it is relatively easy (but also reasonably expensive) to select personnel for occupational exposure, to control the health status periodically during employment and to analyse epidemiologically the data available from medical examinations.

It is our feeling that in the case of occupational exposure major importance should be given to medical/epidemiological data (including possible delayed health hazards!!), while theoretical considerations and experimental observations should have only adjuvant application. In other words, the medical effects should be decisive for evaluating whether or not the real levels of occupational exposure are safe. This assumes however the correct evaluation of individual exposures during relatively long periods of time (e.g. records of exposure, measurements of power density and characterization of EMFs in the workplace), and the specialistic and periodical medical examinations of subjects that for various reasons quit their employment in the EMF environment. Application of modern epidemiological techniques to evaluate the data, and use of valid control groups are further crucial points for obtaining conclusive results. The best solution seems to be to control the whole population living under similar social conditions (e.g., the whole population of career military personnel), to differentiate and well define subpopulations exposed and non-exposed to EMFs and to perform prospective epidemiological studies during 10-20 years. A variety of other environmental and occupational factors (e.g., consumption of tobacco and alcohol, chemical pollutants, occupational and social stressors) should also be considered in the analysis. This is the ideal (but expensive) solution and in the present literature we were not able to find a single study fulfilling these requirements.

In this situation, as in the case of environmental exposure, the occupational safety levels have to be established arbitrarily, on the basis of scarce and fragmentary medical data with an overestimation of results of experimental investigations on animals. In the USSR and Eastern European countries the occupational safety levels are set as low as it is economically acceptable. The selective and periodical medical examinations of personnel are aimed at assessing the health risk at the real conditions of exposure, making eventually possible the gradual increase of the safety levels in accordance with the health status of personnel. As there exists a general agreement that for RF/MWs no significant harmful effects occur at resonant frequencies below 1 mW/cm^2, except for transient adaptation phenomena that cannot be regarded as health hazards, it seems logical to accept this level as suitable for a continuous occupational exposure during a working day (8 hours). Lowering the exposure time and using individual protection devices (suits, glasses), it seems safe to exceed the MW field intensity of about a factor of 10 (10 mW/cm^2), which is a value that does not yet result under experimental conditions in hyperthermia of the whole body.

In the recent Soviet guidelines for occupational exposure standards to MWs (Savin et al., 1983; Akoev, 1986; Shandala, 1986) it is stated that the 0.1 mW/cm^2 for 2-hour daily exposure (a "safety factor" of 10 with respect to the 1 mW/cm^2 value) and the energetic load resulting from field power density of 0.2 mW/cm^2/hour should be adviced as occupational safety levels for 8-hour daily exposure. Both the 1 mW/cm^2 limit as power density not leading to harmful effects and the "safety factor" of 10, being a certain echo of our habit of using the decimal system, have been set by the above Soviet authors arbitrarily, but at the present state of art nothing better can be adviced. Certainly, as in the case of environmental (public) safety levels of EMFs, the above occupational exposure criteria are subject to amendments in the future.

EMF SAFETY CRITERIA IN POLAND AND COMECON COUNTRIES

In Tables 3 and 4 the environmental (public) and occupational safety levels of EMFs currently in force in Poland are shown.

Table 3. Electromagnetic safety levels for the general public in Poland

Frequency	Quantities and units	Minimum value	Maximum value
50 Hz	E (kV/m)	1	10
1-100 kHz	E (V/m)	-	-
	H (A/m)		
0.1-10 MHz	E (V/m)	5	20
	H (A/m)	-	-
10-300 MHz	E (V/m)	2	7
	H (A/m)	-	-
300 MHz-300 GHz	P (W/m^2) Stationary antenna	0.025	0.1
	P (W/m^2) Rotating antenna	0.25	1

Table 4. Occupational electromagnetic safety levels in Poland

Frequency	Quantities and units	Safety zone	Intermediate zone	Hazardous zone	Time limits (h)	Max. value
50 Hz	E (kV/m)	1	10	15	-	15
1-100 kHz	E (V/m)	100*	200*	?	-	200?
	H (A/m)	10*	20*	?	-	-
0.1-10 MHz	E (V/m)	20	70	1,000	560/E	1,000
	H (A/m)	2	10	250!!	560/E	250
10-300 MHz	E (V/m)	7	20	300	$3,200/E^2$	300
	H (A/m)	-	-	-	$3,200/E^2$	-
300 MHz-300 GHz	P (W/m²) Stationary antenna	0.1	2	100	$32/P^2$	100
	P (W/m²) Rotating antenna	1	10	100	$800/P^2$	100

* Pending (1987)

In 1972, when the currently operating Act on safety criteria for MWs was issued, the concept of exposure zones (safety, intermediate, hazardous and dangerous) was first applied for occupational exposure; for public exposure only two zones (1st, unlimited and 2nd, limited) were set. Furthermore, upper and lower limits and defined limitations of exposure in each zone were identified.

The concept of unlimited and limited exposure zones for environmental and occupational conditions (Fig. 1) is an original Polish solution to EMF safety criteria, being applied in practice for about 20 years. This concept is not yet applied in other Eastern European countries, although its introduction is discussed for the proposed uniform EMF safety levels for Comecon countries.

The above concept of unlimited and limited exposure zones is a compromise between the desire to set safety levels as low as it is possible (due to the earlier discussed fragmentary knowledge of biological effects of EMFs and interaction of EMFs with other harmful factors) and the economic costs of the above safety levels. For unlimited exposure a higher "safety factor" is introduced that should protect the population against unsuspected or presently unknown health hazards. Based on a 20-year experience it became evident that the establishment of the above zones facilitates the marking of regions with unlimited dwelling of human subjects in EMFs.

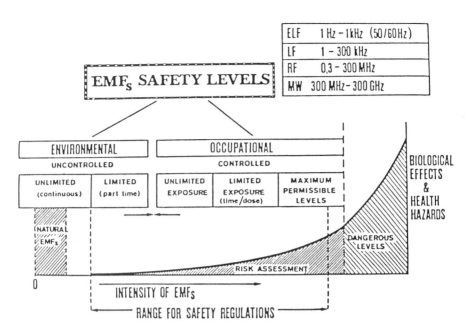

Fig. 1. The concept of EMF safety levels, with zones of unlimited and limited exposure, developed and introduced in Poland. Different safety levels should be established for 4 ranges of EMFs (ELF, LF, RF and MW) with different criteria for environmental (public) and occupational exposures. For both kinds of exposures a maximum level for unlimited (continuous) exposure should be esstablished, as well as maximum permissible levels for limited (part time) exposure. Explanation in the text.

The safety levels presently in force in Poland have been elaborated relatively long ago. The limits were established in 1972 for occupational exposure to MW fields, in 1977 for occupational exposure to RF fields, and in 1980 for environmental exposure to EMFs, respectively. From that time only slight changes in the interpretation of safety zones and enforcing regulations were introduced, but levels were not changed. Thus, it is generally accepted by experts in this country that the operating safety regulations do not meet criteria of the present knowledge and there exists a need for amendment. However, as at present there is a tendency to unify the EMF safety criteria in the USSR and Eastern European countries and an international working group of experts from the Comecon countries is engaged in this issue, it was decided not to change the Polish regulations regarding this matter (Tables 3 and 4) until the uniform criteria will be definitely accepted.

Safety Levels for Environmental (Public) Exposure

Safety regulations for public exposure to EMFs were established in Poland by the Act enforced in 1980. This Act covers power frequency (50 Hz in this country), RFs (divided into the ranges of 0.1-10 MHz and 10-300 MHz) and MWs (300 MHz-300 GHz).

The regulations for ELFs and LFs (1-100 kHz) are still pending. For each of the above frequencies the minimum and maximum environmental safety levels were arbitrarily established (Table 3) and two exposures zones (1st and 2nd) with field strengths between the above minimum and maximum levels were delimited. It is assumed that below the minimum levels (1 kV/m for power frequency, 5 V/m for 0.1-10 MHz, 2 V/m for 10-300 MHz, 0.025 W/m² for stationary, and 0.25 W/m² for rotating 300 MHz-300 GHz fields) there is no need for controlling the exposure and it is believed that EMFs below these levels are safe and do not cause any effects, including the delayed health disorders.

On the other side, maximum EMF strengths allowed for public exposure (Table 3) - 10 kV/m for power frequency, 20 V/m for 0.1-10 MHz, 7 V/m for 10-300 MHz, 0.1 W/m² for stationary and 1 W/m² for rotating 300 MHz-300 GHz fields - are also considered being safe, but not advisable for permanent, uncontrolled exposure (e.g. location of buildings, hospitals, schools, etc.).

At field intensities between minimum and maximum values stated by the Act (2nd zone) there are in force certain limitations (see below) and it is assumed that the exposure in the 2nd zone will be in practice limited to few hours per day, although still uncontrolled. The 1980 Act, being in force at present in Poland, defines all regions above the established maximum environmental safety level (Table 3) as prohibited for uncontrolled exposure (trespassing is forbidden, except for personnel authorized to work with EMF sources and medically selected).

In the 2nd exposure zone (between minimum and maximum environmental EMF levels) a temporary presence (connected with management, touristic or recreative activities and agriculture) of human beings is allowed, even if not quantified. In this zone to localize living quarters and buildings qualified for special protection (hospitals, schools, kindergartens, infants' day nurseries, elementary schools, sanatoria, etc.) is consequently not allowed, and it is adviced that it should be used for parks, recreative areas, and/or agricultural farms.

As it was mentioned before, the concept of limited public exposure in the 2nd zone was introduced in Poland mainly for economical purposes with the aim of avoiding the exclusion of large areas close to EMF sources from human activity in view of the very low minimum levels for uncontrolled environmental exposure. At present there is a tendency in Poland toward the amendment of the minimum and maximum EMF safety levels of public exposure. One of the proposals is presented in Table 5.

In comparison to the present safety levels in Poland, the proposal elevates minimum and maximum values in the MW range (300 MHz-300 GHz), unifies the levels for the whole RF range (0.1-300 MHz) and lowers the limits for power frequency (50 Hz).

The will to reach the unification of EMF safety regulations in Comecon countries led to a proposal elaborated by Soviet experts (Table 6). This proposal does not follow the rule of two exposure zones (unlimited and limited) and sets only one environmental level (maximum) with division of the 30 kHz-300 GHz frequency range into 5 subranges (Table 6), while in the Polish proposal there are only 3 subranges (1-100 kHz, 100 kHz-300 MHz, 300 MHz-300 GHz).

Thus, the above Soviet proposal is still a subject for criticism by Polish experts, as it would be desirable from a practical point of view to establish only one integrated safety level for the whole range of RF/MW radiation due to the known complexity of EMFs in the environment.

As the single safety level for MWs and RFs is still a controversial issue, we propose at present for public exposure 5 V/m for 0.1-300 MHz and 0.5 W/m² (50 µW/cm²) for 300 MHz-300 GHz.

Table 5. Proposed electromagnetic field safety levels in Poland

Frequency	Quantities and units	Public Min.	Public Max.	Occupational Basic	Occupational Additional*	Occupational Maximum
50 Hz	E (kV/m)	0.5	5	10	15	25
1-100 kHz	E (V/m)	50	100	100	1,000	1,000
	H (A/m)	0.5	1	10	100	100
0.1-10 MHz	E (V/m)	5	15	15	150	300
	H (A/m)	-	-	2	4	10
10-300 MHz	E (V/m)	5	15	15	75	150
	H (A/m)	-	-	1	2	3
300 MHz-300 GHz	P (W/m^2) Stationary antenna	0.1	2	2	10	100
	P (W/m^2) Rotating antenna	0.5	2	2	10	100

* Additional exposure - 1 h daily

Table 6. Proposed electromagnetic field safety levels in Comecon countries

Frequency Range	Maximum value
30-300 Hz	5 kV/m
30-300 kHz	25 V/m
0.3-3 MHz	15 V/m
3-30 MHz	10 V/m
30-300 MHz	3 V/m
300 MHz-300 GHz	0.1 W/m^2 (10 µW/cm^2)

Safety Levels for Occupational Exposure

The safety criteria for occupational exposure operating at present in Poland (Table 4) cover only the RFs (0.1-300 MHz) and MWs (300 MHz-300 GHz) and were elaborated and published in 1977 and 1972, respectively. Maximum strenghts of RF/MW fields allowed for occupational exposure have been established and the following exposure zones have been differentiated:

- safety zone: uncontrolled exposure of personnel, intensities below maximum environmental safety level;

- intermediate zone: personnel selected for employment at EMF-generating devices on the basis of medical examinations and fit for work in the EMF environment may dwell in this zone without time limits during working day (8-12 hours, depending on type of work or service);

- hazardous zone: personnel (selected, as above) may dwell in this zone only for a limited time (Table 4) and there is no need for individual protection devices (suits, glasses, screens, etc.);

- dangerous zone: trained and selected personnel may dwell only temporarily in this zone with individual protecting devices limiting the exposure levels to at least those occurring in the hazardous zone.

On the basis of field measurements in working places in the vicinity of EMF-generating devices the above 4 zones should be delimited and signed.

The safety restrictions in force in Poland state that in employment of personnel in the EMF environment four basic principles have to be obeyed:

1. Selective, general, ophtalmological and neurological medical examination before employment at EMF-generating devices with precise criteria for fitness to occupational exposure.

2. Periodical (every 1-3 years, depending on type of work or service) medical examinations with individual records (each employee has an individual booklet with records of exposure and medical examinations).

3. Transferring of personnel showing certain health disorders (listed in the instructions for RF/MW safety, including for example changes in eye lenses, liability of blood pressure, vegetative neuroses, EEG changes).

4. Time limits for permanence in the hazardous zone and obligatory use of protective suits and glasses in the dangerous zone.

The medical records of personnel exposed occupationally to RF/MW radiation in Poland, at least those of military career personnel, did not report till now alarming symptoms that may be related to exposures. The number of workers transferred from work characterized by the use and/or repair of RF/MW-generating devices, increases only unsignificantly the morbidity observed in personnel not exposed to RF and/or MW radiation. Some confusion was however generated recently regarding the operating restrictions of EMF safety criteria by the retrospective epidemiological analysis of cancer morbidity among career military personnel in Poland during the decade 1971-1980 (Szmigielski et al., 1988; Szmigielski and Gil, this book). In general, this study has documented that the subpopulation of the personnel exposed occupationally to RF/MW suffered about three times more frequently from various neoplastic diseases with preference for certain hemopoietic and lymphatic neoplastic syndromes as well as alimentary tract and skin neoplasms, compared with non-exposed personnel. The risk of neoplastic disease in the tested group has shown a close relation with period of exposure to RF/MW fields. More detailed analysis and discussion of the data (Szmigielski et al., 1988) suggest that in this case the RF/MW exposure accelerated the clinical appearance of neoplasms that without the esposure would develop few years later (postulated tumor-promoting, but not carcinogenic potency of RF/MW radiation). As this is the first and only epidemiological study on cancer morbidity in personnel occupationally exposed to RF/MWs, the problem needs both confirmation and further investigation, until it may affect the occupational safety criteria. The prospective epidemiological study, planned for 1986-1990 and now in progress, should bring about new and more convincing information on cancer risk related to RF/MW occupational exposure. Thus, it is the present authors' feeling that the occupational RF/MW safety criteria and levels should not be changed, until the problem of cancer risk will become clearer.

In general, the present occupational safety regulations in Poland are considered by our experts as providing sufficient health protection of the exposed personnel and can be relaxed in the future. It is however stressed that the magnetic field strengths (H, expressed in A/m) for the frequency range of 100 kHz-10 MHz (Table 4) were set too high and need reconsideration. In the recent proposal for amendments (Table 5) much lower magnetic field strengths for the frequency range of 100 kHz-10 MHz were introduced.

In 1984, and later in 1987, the group of research workers of the Institute of Labour Medicine in Lodz, Poland, headed by Professor H. Mikolajczyk (Mikolajczyk, 1987) proposed new criteria for occupational exposure for the whole EMF spectrum. This proposal was a subject of discussion among the Polish experts and was slightly modified (Table 5). Due to this amendment, basic exposures for not more than 1 hour per day, as well as maximum intensities of EMFs permissible in occupational exposure without special protection were established. Comparing the present safety levels in Poland (Table 4) with those proposed (Table 5) one can only see slight modifications of maximum permissible levels and increase of the values for safety and intermediate zones. The above proposal covers also the 1-100 kHz range (LFs), being at present a subject of consideration by two groups of Polish experts. However, the data from medical examination of workers using the 10-50 kHz generators are still not conclusive, as only about 200 subjects were included in the analysis and thus the proposals for 1-100 kHz (Tables 4 and 5) should be considered as general guidelines being a subject of more accurate definition in the future.

CONCLUSIONS

Safety criteria of public and occupational exposure to EMFs are based in Poland, the USSR and other Eastern European countries on analysis of data from medical examinations of personnel and epidemiological observations of large groups of population working or living in the EMF environment. Experimental data from irradiated animals, as well as theoretical considerations and calculations of power absorption, current flow, etc. serve only as supportive guides to set the safety standards below the possible hazards. Basing safety criteria on medical and epidemiological data from personnel exposed to harmful factors is a philosophy different from that followed in Western European countries and in the USA and needs to set very low limits at the beginning and a future relaxation when it appears clearly that they do not increase risk of diseases.

Biological effects of long-term exposure to weak EMFs are generally transient, inconsistent and difficult to replicate and interpret;

nevertheless a variety of non-specific behavioural, neurological, immunological and endocrine reactions were reported in experimental animals (Elder and Cahill, 1984).

Shandala (1986) claims that the immunological status of animals and human beings appears to be the most susceptible indicator of harmful effects of all environmental factors, including EMFs, and changes in immunological reactivity are among the earliest effects observed in the organism and thus may be used as indicators of prepathologic changes in other organs and systems. Unfortunately, this is only a general statement and the authors are not able to provide convincing evidences that it is true in the case of EMFs.

On the basis of experiments performed by Soviet scientists, Shandala (1986) claims that long-term exposure to weak MW fields results in the lowering of non-specific immune reactivity (e.g. phagocytosis, complement titers, etc.) that leads to decrease of antibacterial, antiviral and antineoplastic resistance of the organism. At the same time stimulation of autoimmune reactions with appearance of autoantibodies have been observed (Shandala, 1986). The above effects were observed in mice exposed to 2,450 MHz MWs at power densities of 50-500 $\mu W/cm^2$ and thus these intensities were recognized by Soviet authors as harmful for the organism.

It is difficult to accept all these findings, due to the fact that the conditions of the cited experiments were not described in sufficient detail, but it may be concluded that more attention should be paid to reactivity of the immune system to EMFs that till now was investigated only fragmentarily.

It is worth noting here that not all changes in the function of organs and systems (both in experimental animals and in medical examinations of human subjects) may be recognized as indicators of harmful effects of the tested factor. A known property of living organisms is their ability of adaptation to a variety of environmental factors and this occurs also in case of EMFs among non-specific biological stressors. Thus, in terms of theoretical consideration, the biological effects of weak EMFs should be divided into physiological adaptation, compensatory reactions, reparative regeneration and pathological changes leading to organic or systemic diseases. Review of the available literature and personal experience of the present authors suggest that exposure to low EMFs leads only to adaptative and compensatory reactions, mostly from the central nervous, endocrine and immune systems.

By definition, permissible levels for environmental factors, accepted in environmental and occupational hygiene, are those corresponding to the intensity of any factor (e.g. EMFs) which after action on human organisms (periodically or during the whole life, directly or in combination with other ecological factors) does not result in the development of somatic or psychical diseases (including temporary compensated) or in other health disorders which exceed the range of adaptative reactions that may be diagnosed, with methods presently available, either directly during exposure or later on in life or in the following generations.

In case of EMFs it is not possible at the present state of art to conform safety criteria to the above definition and any attempt will be only an approximation. It is our strong feeling that in case of uncertainty concerning biological effects and health hazards of environmental and occupational exposure to EMFs, it is to start with low safety levels and more rigorous restrictions (even if the future progress of bioelectromagnetics will prove the restrictions being too rigorous) than underestimate the risk and follow the opposite concept.

REFERENCES

Adey, W. R., 1981, Ionic nonequilibrium phenomena in tissue interactions with electromagnetic fields, in: "Biological Effects of Nonionizing Radiation," K. H. Illinger, ed., ACS Series 157, American Chemical Society, Washington.

Akoev, I. G., 1986, Principles of biological risks and safety criteria of electromagnetic radiations, in: "Biological Effects of Electromagnetic Fields. Problems of Their Use and Safety," I. G. Akoev, ed., USSR Academy of Sciences, Pushtshino (In Russian).

Akoev, I. G., Alekseev, S. I., Tjazelov, W. W., and Formenko, B. S., 1986, Primary mechanisms of action of radiofrequency radiation, in: "Biological Effects of Electromagnetic Fields. Problems of Their Use and Safety," I. G. Akoev, ed., USSR Academy of Sciences, Pushtshino (In Russian).

Baranski, S. and Czerski, P., 1976, "Biological Effects of Microwave," Dowden, Hutchinson and Ross, Stroudsburg.

Byus, O. V., Lundak, R. L., Fletcher, R. M., and Adey, W. R., 1984, Alteration in protein kinase activity following exposure of cultured human lymphocytes to modulated microwave fields, Bioelectromagnetics, 5:341.

Choy, R., Monro, J., and Smith, C., 1987, Electrical sensitivity in allergy patients, Clinical Ecology, 4:93.

Davidov, B. I., Tichonczuk, W. S., and Antipov, W. W., 1984, "Biological Action, Safety and Protection Against Elecctromagnetic Radiations," Elektroatomizdat, Moscow (In Russian).

Elder, J.E., and Cahill, D. F., eds., 1984, "Biological Effects of Radiofrequency Radiation,", EPA Report, 600/8-83-026F, USEPA, Research Triangle Park.

Gordon, Z. A., 1974, "Biological effects of Extremely High Frequency Electromagnetic Fields," Medicina, Moscow. (In Russian).

Marino, A. A., 1988, Environmental electromagnetic fields and public health, in: "Modern Bioelectricity," A. A. Marino, ed., M. Dekker Inc., New York and Basel.

Mikolajczyk, A., 1987, "Experimental and medical investigations of effects caused by 1-100 kHz electromagnetic fields and guidelines for safety levels," Final Report of the CILB Grant, Institute of Labour Protection, Lodz (In Polish).

Minin, B. A., 1974, "Ultra-High Frequencies and Human Safety," Elektroatomizdat, Moscow (In Russian).

Savin, B. M., Nikonova, K. W., Lobanova, E. A., Sadczikova, M. N., and Lobed, E. K., 1983, Novelties in Safety Standards of EM Radiation of the Microwave Range, Gigiena Truda, 3:1 (In Russian).

Shandala, M. G., 1986, Methodologic problems of hygienic standardization of nonionizing electromagnetic radiations, in: "Biological Effects of Electromagnetic Fields. Problems of Their Use and Safety," I. G. Akoev, ed., USSR Academy of Sciences, Pushtshino (In Russian).

Szmigielski, S., Bielec, M., Lipski, S., and Sokolska, G., 1988, Immunologic and cancer-related aspects of exposure to low-level microwave and radiofrequency fields, in: "Modern Bioelectricity," A. A. Marino, ed., M. Dekker Inc., New York and Basel.

EXISTING SAFETY STANDARDS FOR HIGH VOLTAGE TRANSMISSION LINES

Martino Grandolfo, and Paolo Vecchia

National Institute of Health, Physics Laboratory
and INFN-Sezione Sanita', Rome, Italy

INTRODUCTION

The ever larger diffusion of electric appliances and the consequent demand for electric power have greatly increased in the last years the awareness, both in the scientific community and among the general public, of health risks of electric and magnetic fields at extremely low frequencies (ELF). In fact, the field strengths which are experienced in the proximity of electric equipment or near high-voltage transmission lines may reach such levels that the possibility of adverse health effects cannot be excluded (WHO, 1984; WHO, 1987).

Concern has particularly raised about power lines, also due to the present tendency to increase the operation voltage up to and above one megavolt. That led in some cases to strong opposition of the population against the installation of transmission facilities, which delayed or even prevented their licensing. Although the number of such cases is limited, the request for assurance about health risks by workers and by the public has increased to the point that some regulatory agencies have confronted with the problem of issuing standards on the exposure levels to 50/60 Hz electric and magnetic fields.

According to a generally accepted definition (WHO, 1984; Repacholi, 1985), a standard is a general term indicating a set of specifications or rules to promote the safety of an individual or group of people. Standards can be divided into regulations and guidelines. A regulation is a mandatory standard promulgated under a legal statute, whereas a guideline is in general a set of recommendations issued for guidance only. A standard can specify any kind of rule to be complied with in order to reduce health risks within acceptable limits. In the case of electromagnetic fields, such specifications include field limits, i.e. maximum levels that the electric or magnetic field may attain in specific points or regions; exposure limits, i.e. maximum levels to which the exposure of the whole body or part of it is allowed; interdiction or limitation of access to specific areas; rules for the siting of an industrial facility; details on the performance, construction, design or functioning of a device.

A limit should be based on the ascertainment of a threshold value for adverse health effects; in the absence of any apparent threshold, a level of acceptable risk should be determined, based on a risk-benefit analysis, which in turn requires some correspondance to be established between different exposure levels and biological or health effects. The definition of such limits are the basis for promulgating protection standards.

Obviously, the enforcement of mandatory regulations is justified when biological and health effects of noxious agents are ascertained, and threshold values, or some relation to the exposure, can be determined for these effects. In the absence of such data, the adoption of guidelines is more appropriate, often as an interim measure until sufficient information about adverse effects becomes available.

As pointed out by several authors (Grandolfo et al., 1985; Tenforde and Kaune, 1987), data on biological effects of ELF fields and their mechanisms of interaction with cellular and tissue systems are at present controversial and insufficient to derive any clear exposure-effect relationship. That is the reason why only a few standards have so far been issued, and mainly as guidelines.

A standard is often issued in response to solicitation by groups of workers or by the public opinion. That derives in turn from concern about health hazards, which is in several cases exaggerated by alarmist reports published in the press or presented by the media. At present, the greatest concern is for high-voltage transmission lines, and sometimes it gives rise to protests and strong opposition towards the installation of new facilities.

As already mentioned, the intervention of objectors at different levels of the decision-making bodies may cause even long delays in the licensing procedure. On the contrary, protests do not significantly affect the routine operation of transmission networks, because they usually cease after the facility is installed.

Taking into account the economic and social cost of the above mentioned delays, and sometimes of legal actions, it may be useful also for the electric utilities that some limit values be defined, provided they are technically reasonable and scientifically valid. This opinion has been recently expressed, for example, by the Committee for Medical Studies of the International Union of Producers and Distributors of Electrical Energy (UNIPEDE, 1985).

In this paper, a review of the existing exposure standards for 50/60 Hz electric and magnetic fields is presented, with particular attention paid to the scientific data on which they are based.

STANDARDS ISSUED IN DIFFERENT COUNTRIES

General overview

Although the use of electrical appliances has been widespread for several decades, and the exposure to electric fields has continuously increased, protection problems had been largely ignored until the end of the sixties, when the early studies of Soviet medical teams (Asanova and Rakov, 1966; Sazonova, 1967) were published, reporting a number of subjective complaints by high-voltage substation workers.

The findings of Soviet investigators stimulated a large number of studies on a variety of possible effects of electric fields on biological systems. The results are still controversial; anyway, they show that a biological response, if any, can be observed at levels of the electric field which can be practically experienced only in the vicinity of high-voltage transmission lines. Therefore, standards which have been issued or are being evaluated deal almost exclusively with power lines and related facilities, such as high-voltage conversion substations.

A general overview of the present status of regulations, and of problems associated with electric and magnetic fields from power transmission systems in industrialized countries, is given by the results of an international survey performed by the Study Committee 36 of the International Conference on Large High Voltage Electric Systems (CIGRE, 1986).

The aim of the survey was to get general information about: experience gained in developed countries with large high-voltage networks; attitude of public opinion; knowledge of fields generated by power lines and their physical effects; research either in progress or planned on field effects; legal matters such as regulations, disputes, and legal actions taken by opposers. For this purpose, a questionnaire was distributed in all the member countries of the Study Committee. These countries are twenty-one in all, with transmission lines operating at various voltage levels, as indicated in Table 1. Table 1 also lists the respective route lengths.

The results of the survey show that, even in developed countries with extended electric networks, the levels of fields generated by the lines and the related problems are not completely appreciated. For example, only 17 countries have the capability to make calculations of electric fields, and not all have access to instruments for its experimental measurement.

As far as exposure standards are concerned, almost half (10 out of 21) of the countries have no regulations or guidelines giving limit values for the electric field under lines. For the others, 5 out of 11 have limits imposed or recommended by some public agency or department, whereas in the remaining six countries some protection is assured at the time of the design of a new line, by conforming the project to design principles established by the utilities themselves.

Due to the relatively scarce knowledge about biological effects, and to the different degree of appreciation of the health hazard, the reasons given for limits may be different. In some standards, limits aim at reducing the discharge current from large objects; in others at avoiding microshocks and similar unpleasant effects; in others, they just tend to restrict long-term exposure to high fields.

In general, limits which are presently adopted in a given country, or region, are such as to pose no significant constraint on the present design practice, taking into account the geographic, economic, and demographic characteristics of the involved areas (CIGRE, 1986).

The problem of possible health effects of ELF magnetic fields has so far received much less attention than electric fields. A thorough review of the scientific literature on biological effects of ELF magnetic fields is given in two Environmental Health Criteria Documents developed by the IRPA/INIRC in cooperation with the World Health Organization (WHO, 1984; WHO, 1987) and in Bernhardt (1986).

The current knowledge of biological effects seems indeed inadequate to confirm or disprove any possible health hazards. It would therefore be premature to set mandatory standards in this area. Nevertheless, in recent years several authors have reported possible correlations between cancer and prolonged presence near electric wiring characterized by high current intensities, both in houses and in workplaces. Some of the

Table 1. Operation voltage and transmission route length of electric networks in countries participating in the CIGRE survey

Country	Voltage (kV)	Length (km)	Country	Voltage (kV)	Length (km)
Australia	500	1,420	Netherlands	400	574
	330	4,800	Norway	420	1,140
	275	3,640		300	3,570
Belgium	400	802	Poland	750	114
Brazil	750	570		400	2,140
	500	9,260	South Africa	400	8,926
	345	6,800		275	5,965
Canada	750	9,600	Spain	380	8,496
	500	9,060	Sweden	400	8,957
	345	7,600	Switzerland	400	1,000
Czechoslovakia	400	4,500	UK	275	1,693
Denmark	400	365	USA	765	3,100
F. R. Germany	380	10,250		500	32,000
Finland	400	3,200		345	45,000
France	400	9,013	USSR	1,150	800
Italy	420	5,300		750	3,800
Japan	500	3,456		500	34,000
	275	6,776		330	27,000

authors suggested that the magnetic field, rather than the electric field, may act as a causal or at least as a promoting factor. Although these findings have been contradicted by the results of other authors, which show no apparent difference between exposed and control groups, the above mentioned papers have raised interest on health effects of ac magnetic fields, and a number of well designed surveys are being planned.

The term magnetic field is often applied to both the magnetic flux density (B) and the magnetic field strength (H). Some guidelines express exposure limits in terms of H, others prefer B. The units of H are amperes per meter (A/m), those of B are tesla (T). Since biological materials are mostly nonmagnetic, permeability is usually not an important factor in bioelectromagnetic interactions. In this case, 1 A/m corresponds to about 1.3 µT and, conversely, 1 µT to about 0.77 A/m.

So far, only the USSR has issued an official regulation limiting exposures to ELF magnetic fields. Some agencies, however, are examining the possibility of establishing standards on ELF magnetic fields. In two countries, namely FRG and UK, a proposal for limits has been published.

In the following, standards are examined in more detail, limited to those issued by national departments, public agencies or international organizations independent of the electric utilities.

Australia

Each of the six States costituting the Australian Federation has its own electricity authority. None has promulgated legal standards for the protection of workers or population against the effects of ELF fields.

Only the two largest States, New South Wales (NSW) and Victoria, have guidelines for the costruction of power lines. It is to be noted that these are the only States having lines operating at 330 and 500 kV, whereas in the rest of the country the highest voltage is 275 kV. Anyway, recent guidelines of Victoria recommend field strength limits also for 220 kV lines (Smith, 1985). The recommended limits in the two States are shown in Table 2. As an additional protective measure, the height of 500 kV lines at crossing with main roads is raised to 15 m to limit the discharge current below the let-go threshold. The principle of right-of-way (RoW) also applies, with minimum widths of 60 m for 330 kV lines and 60-70 m for 500 kV lines. Inside this zone, buildings and full-time activities are prohibited. Also structures which could lead to activity above ground level are not permitted.

As far as workers are concerned, design principles for substations are set, based on which the maximum field strength at ground level is 10 kV/m in the vicinity of extra-high-voltage (EHV) equipment or overhead busbars, and less than 5 kV/m in normal access areas.

In 1973, Victoria established an interim set of restrictions for work in the vicinity of 500 kV apparatus (Smith, 1985). In particular, limits were set for the exposure times, based on the expected body current. Pending the results of investigations being carried out on an international basis, the exposure times listed in Table 3 shall not be exceeded when the personnel are in the vicinity of live 500 kV apparatus.

Table 2. Recommended limits for electric field strengths in Australia (Smith, 1985)

State	Line voltage (kV)	Max. E-Field (kV/m) in RoW	Max. E-Field (kV/m) edge RoW	Comments
NSW	500	-	2	---
Victoria	500	10	2	Exceptional conditions
		5	2	Normal conditions
	220	5	2	Exceptional
		2.5	1	Normal conditions

Table 3. Limits for exposure times for workers at 500 kV apparatus in Victoria (Smith, 1985)

Body current (µA)	Permitted exposure time
> 320	Not permitted
320 down to 80	10 min increasing to 4 h per day
< 80	Continuous

Note: 80 µA is the body current expected in an unshielded man of average size, subjected to a field of 5 kV/m measured at ground level.

The rationale for the guidelines is given by Johnson et al (1976), and is also based on the fact that the proposed limits are generally accepted in other countries. The more recent limits of Victoria for 220 kV lines have been suggested in the 1982/83 *Inquiry into Transmission Lines Serving Melbourne,* performed by the Natural Resource and Environment Committee for the Power Authority of the State, and are based on a world research into biological effects of transmission lines.

Czechoslovakia

Some information about present regulations in Czechoslovakia is given by Kabrhel (1985).

For presently operated 400 kV power lines, a right-of-way area is required, up to a distance of 25 m from the outer phase. Provided safe conditions are ensured, agricolture is allowed within the RoW. Building or living in this area is prohibited. Cars and agricoltural machines are not allowed to stop below the line.

For the population the Czechoslovack standard 33-2040 *Protection against the Influence of Electric Field in the Vicinity of Electric Transmission Systems of 750 kV and above* establishes the limits reported in Table 4.

For workers, the *Guide for Stay of Staff in 400 kV Substations* requires to design new substations as to limit electric field to 15 kV/m. Under these circumstances, no further precautions are used.

Table 4. Electric field limits in Czechoslovakia (Kabrhel, 1985)

Maximum E-field (kV/m)	Comments
15	Within right-of-way
10	At main road crossings

Federal Republic of Germany

In the Federal Republic of Germany (FRG, 1986), a draft of a national standard for the protection of persons against hazards from electromagnetic fields in the frequency range from 0 Hz to 3,000 GHz was published in August 1986, and submitted to the public for objections and suggestions. The part of the standard concerning the frequency range between 0 Hz and 30 MHz has been approved now (FRG 1989).

In the frequency interval 0 Hz-30 kHz, which includes the whole range of extremely low frequencies, the limit for the rms value of the electric field strength is given by the following expression:

$E = a/f^b$

where E is the limit value in V/m, f is the frequency in hertz, whereas a and b are constants whose value is given in Table 5.

The above formula allows the determination of a limit for the effective electric field. The proposed standard also sets a limit for the peak amplitude, which is at each frequency 1.5 times the limit for the rms field strength. At the single frequency of 50 Hz, their values are practically equal to 20 and 30 kV/m, respectively.

The definition of the limit value is based on the unperturbed homogeneous field. In the case of inhomogeneous electric field, an equivalent homogeneous field must be calculated, through measurement procedures that allow a comparison with the total current dispersed through the human body in the unperturbed field.

Exposure to electric fields below the limits is allowed for unlimited time. For short-time exposures (up to 2 hours per working day), exposure to field strengths up to 1.5 times the limit value is allowed. The possibility that some people feel oppressed by electric fields of intensity of 2 to 10 kV/m, or may experience unpleasant electrical discharges when touching charged objects or other people in an electric field, is recognized, but no scientific data exist at present, which allow to relate these sensations to health effects.

Such considerations, as well as the scientific basis of the current density concept which is partially used are in papers referenced in the document (Bernhardt et al., 1983; Bernhardt, 1984).

The standard sets limits also for low-frequency magnetic fields. The basic criteria are essentially the same. Also for magnetic fields, a maximum permissible level is defined in the frequency range interval 0 Hz-30 kHz, through the expression:

$$H = c/f^d$$

where H is the limit value in A/m, f is the frequency in hertz, whereas c and d are constants, whose value is given in Table 6.

The above formula allows the determination of a limit for the effective magnetic field. The proposed standard also sets a limit for the peak amplitude, which is at each frequency 1.5 times the limit for the rms strength.

At the single frequency of 50 Hz, their values are practically equal to 4,000 and 6,000 A/m, respectively.

Table 5. Values of the parameters used in the calculation of the ELF electric field limits in FRG (1986, 1989)

Frequency (Hz)	a	b
0 - 10	40,000	0
10 - 30,000	102,850	0.4101

Table 6. Values of the parameters used in the calculation
of the ELF magnetic field limits in FRG (1989)

Frequency (Hz)	c	d
0 - 2	under consideration	0
2 - 30,000	21,593	0.4325

The definition of the limit value is based on the case of a homogeneous field. In the case of inhomogeneous magnetic field, an average is to be calculated over a circular area of 100 cm^2.

The above limits are intended for unlimited exposures of the whole body. For prolonged exposures of the extremities (hands, arms, feet, and legs) a value up to 10 times the limit is allowed. For short times (up to 5 minutes per hour), exposure to fields of strength up to 1.5 times the limit value is permitted. The scientific basis of the current density concept which is partially used are given by Bernhardt (1986). Effects of currents generated in tissues have been considered, whereas effects which are only related to the perception of the field, but do not constitute any health hazard, such as magnetophosphenes, have been disregarded.

Japan

In Japan, all electrical equipment are subject to the *Technical Standards of Electrical Facilities* ordinance of the Ministry of International Trade and Industry, issued in 1973. This ordinance (Yasui, 1985) includes some specifications aimed at the safety and protection of the public.

The construction of high-voltage overhead lines is prohibited in densely populated areas, with a building coverage of 25% or more. In suburban or rural areas, the installation of a power line is not allowed within a distance less than 3 m from a building. To comply with this standard, electrical companies secure at the design time a right-of-way along the projected line route, where the construction of buildings is prohibited.

The standard also defines a limit for the electric field strength under the line, which must not exceed 3 kV/m at the height of 1 m from ground level. This limit is specified for the purpose of preventing unpleasant sensation on people on account of discharge from metallic umbrellas underneath power lines. Since this uneasy effect only occurs under the specific condition of a person touching the metallic part of the umbrella when passing under the line, the requirement is not applied where people scarcely come and go, as in fields, forests, etc.

The scientific data used to determine 3 kV/m as an average threshold for sensing induced current when a cheek or finger is in contact with a metallic umbrella are reported by the Japan IERE Council (Japan, 1976). As shown in Tables 7 and 8, the perception of field occurs at levels which obviously depend on the individual sensitivity, psycological factors, the characteristics of the umbrella, etc. Anyway, fields below 3 kV/m may be sensed only by some people, whereas fields of 4-5 kV/m may give disturbing sensations. Seven people out of 48 had the perception of the field only above 4 kV/m.

Table 7. Feeling on the cheek for various induced voltages in an umbrella (Japan, 1976)

Electric field strength (kV/m)	Sensation when the cheek is in contact with the umbrella
0.5 - 1	Scarcely sensed
1.5 - 2	May be sensed
2.5 - 3	Slight stimulation may be sensed at the moment of touching
> 4	Fairly well sensed

Table 8. Number of people reporting sensations from umbrellas under 275 kV transmission lines (Japan, 1976)

Electric field strength (kV/m)	Number of people
< 2	0
2.1 - 3	12
3.1 - 4	22
4.1 - 5	7
5.1 - 6	7

The above limits and regulations apply only to the general public. No legal standards exist for workers at substations. As reported by Yasui (1985), possible measures to protect personnel from unpleasant effects of transient discharge due to static induction were examined by a study committee composed of members from the Central Research Institute of the Electric Power Industry, from the electric power companies, and from manifacturers, which established design principles for substations. For example, 275 kV substations are designed in such a way as to limit the field strength above the ground to less than 7 kV/m.

Poland

In Poland (1980), an official standard was enforced in January 1980 to protect people and the environment against non-ionizing electromagnetic fields. In the frame of a general regulation for fields of frequencies up to 300 GHz, limits are also established for the single frequency of 50 Hz.

As a general rule, the maximum power-frequency electric field strength from power lines should not exceed 10 kV/m. To avoid exposure to fields higher than this limit, a protection zone is defined as the area along the line where the electric field exceeds 10 kV/m. In this zone, the presence of anyone other than personnel working on the line is forbidden. A second protection zone is also defined as the area where the field strength ranges between 1 and 10 kV/m. In this latter only the temporary presence of people related to farming, touring, recreation is allowed.

According to the response of Polish authorities to the CIGRE questionnaire (Pilatowicz, 1985), only the second protection zone is of practical importance, since the field levels under power lines do not actually reach 10 kV/m.

For permanent stay, as well as for the construction of residential houses and buildings requiring special protection against field effects, such as hospitals, schools, nurseries, kindergartens, the maximum admissible field strength amounts to 1 kV/m.

For workers, a limit of 15 kV/m is established for the electric field strength in high-voltage substations. In special circumstances, an exposure to fields up to 20 kV/m is allowed during maintenance operations; in this cases, permanent screens must be used.

In our knowledge, no scientific paper giving a rationale for the limits adopted in the Polish standard has so far been published. Some indications are given by Pilatowicz (1985) in the response to the CIGRE questionnaire. Studies performed by medical universities in Poland are mentioned, but not referenced, which show the existence of health effects of electric fields. According to this studies, fields of strength between 16 and 19 kV are threshold stimuli that can lead to organic changes if the action lasts, though an adaptation of the human organism after a certain lapse of time is possible. Electric fields between 10 and 16 kV/m are sub-threshold stimuli, which may cause less perceptible, statistically non-symptomatic deviations in living organisms. Fields of strength below 10 kV/m seem to produce no perceptible changes in living organisms.

United Kingdom

In the spirit of the recommendations of the World Health Organization (WHO, 1981; WHO, 1984), the National Radiological Protection Board (NRPB, 1986) has issued a consultative draft of guidelines for protection against hazard from electromagnetic fields of frequency between 0 Hz and 300 GHz. The document is developed from earlier proposals (NRPB, 1982) and from comments received on them. At this moment, a definitive standard has not yet been promulgated.

As regards low frequency electric fields (up to about 500 kHz), the draft consists of a set of basic limitations on the electric currents and current densities in the body. Corresponding limits for the electric field are also given.

In the advice, a distinction is made between occupational and population exposure.

As far as workers are concerned, the advice is designed to keep exposure below levels at which there is either direct evidence from humans or reasonable evidence from experiments on animals that there will be perceptible effects, due to the direct interaction of the field with the tissues of the human body. Among these effects, which could be stressful if prolonged but may not otherwise be harmful, spark discharges, corona discharges, and involuntary muscle contractions are mentioned.

For the general public, limits are aimed at avoiding even mild perceptible effects, so that exposure levels are kept below those that can be regarded as acceptable for workers occupationally exposed to electromagnetic fields.

Differently from other standards, the NRPB document introduces also a concept of dose through a time averaging and integration regime, which restricts exposures to the highest allowable limits to a given maximum time per day. That is considered as an additional precaution against possible long-term effects, including stress related effects for prolonged exposure at perception levels.

In particular, it is advised that the average exposure of workers at the limiting field values do not exceed 2 hours per day, as averaged over a working week. If working conditions require longer exposures, the levels should be reduced in the ratio 2/T, where T is the exposure duration in hours.

For the population, two limits are set: it is recommended that public access to areas where the electric field strength is more than the upper limit be not allowed, whereas in residential areas the field strength should be less than a lower limit, below which consideration of occupancy and shielding factors is unnecessary.

For areas where the field strength is between the two limits, it is advised that the average exposure time at the limiting values be shorter than 5 hours per day.

For exposures times longer than 5 hours, the field strength should be reduced in the ratio 5/T, where T is the exposure durations in hours.

The recommended limits, both for workers and for members of the public, vary with frequency. Their values at a frequency of 50 Hz are listed in Table 9.

These limits are for rms values of the electric field, and for whole body exposure. For occupational exposures, the electric field strength at the arms, hands, legs, ankles or feet should in no case exceed 30 kV/m.

In the opinion of the Board, the concept of dose has a number of advantages in comparison with the definition of rigid limits. In particular, it conforms to the recommendations of the World Health Organization; it introduces a logical distinction between occupational and population exposure guidelines; it provides greater flexibility, particularly with regard to public exposures.

On the other hand, the Board is aware that no biological evidence supports the details of the proposed regime, and that application of this criterion is more complex than compliance with rigid limits.

Based on the consideration that the production of neoplastic diseases, or birth defects and congenital abnormalities in the offspring of people exposed to electromagnetic fields cannot be excluded, it is also recommended as a general rule, both for workers and the population, that exposure to electromagnetic fields of any frequency be kept as low as reasonably practicable, until it can be shown beyond reasonable doubt that such effects are to be excluded. The Board itself concedes that such a recommendation may be unnecessary, if the effects are of a threshold nature and if limits have been set below these thresholds.

The limits of NRPB have been developed taking into account in particular the Environmental Health Criteria Document of WHO (1984), and the scientific data therein referenced. Anyway, the need for a better assessment of the biological bases underlying the various recommendations is outlined.

Table 9. Limit values of the 50 Hz electric field strength (kV/m) recommended by the NRPB (1986)

Occupational	Population	
	Accessible areas	Residential areas
30	12	2.6

The consultative draft sets limits for magnetic fields as well as for electric fields. As already mentioned, in the low frequency range (up to about 500 kHz), the draft consists of a set of basic limitations on the electric currents and current densities in the body.

From the values of such currents, limits can be derived both for the electric and the magnetic field, based on present knowledge of the dielectric properties of human body.

All the basic considerations that have been exposed for exposure to electric fields, such as the concept of dose, the distinction between occupational and public exposure, the definition of accessible and residential areas with different limits, apply to magnetic fields as well.

The recommended limits at power frequency both for workers and for members of the public, are listed in Table 10.

As for electric fields, in view of possible long-term effects, exposure of workers at the limiting values should be less than 2 hours per day. In the case of longer times, the limit should be reduced by a factor 2/T, being T the exposure time in hours.

For members of the public, exposure at maximum fields in accessible areas should be limited to 5 hours per day, or should the limits reduced in the ratio 5/T.

The standard is aimed at limiting, at power frequency, the current densities in the head and in the trunk of the body to 10 mA/m^2, and the currents in any arm, hand, leg, ankle or foot to 1 mA.

The Board concedes that the scarce available data make it difficult to establish a link between external magnetic fields and internal current densities. However, since the threshold for visual phosphene stimulation by an external magnetic field at 20 Hz is known in terms both of current density and of field strength (Silney, 1985), these values have been taken for normalisation.

Table 10. Limit values of the 50 Hz magnetic field strength (A/m) recommended by the NRPB (1986)

Occupational	Population	
	Accessible areas	Residential areas
1,500	600	138

USA

In the United States, regulations widely differ from a State to another. A review of the existing standards has been performed by Shah (1979); more recent data have been given by Banks (1986) and by Zaffanella (1985), US delegate in the Study Committee 36 of CIGRE, in response to the already mentioned questionnaire.

For the design criteria of new lines, all the States have adopted the National Electric Safety Code (NESC), or some modification of it, for *practical safeguarding of persons from hazards arising from the installation, operation and maintenance of overhead supply and communication lines and their associated equipment*. The code requires the lines to be designed as to limit the discharge current to 5 mA if the largest truck, vehicle or equipment under the line were short-circuited to the ground.

The concept of Right-of-Way is adopted by electric companies in all the country. No house, and no full-time activity, is allowed in the RoW. The width of RoWs depends on the operation voltage of the line and on the policy of the utility. Typical values are given in Table 11, along with typical values of the maximum electric field measured under the lines.

Although most States require preliminary evaluation of public safety and comfort along the area traversed by a proposed line, only a few have recommended guidelines for maximum permissible electric fields. As can be seen from Table 12, the standards are not consistent, since each of the States has different limits inside, or at the boundary of, the Right of Way.

The limits within the RoW are intended to avoid currents above let-go threshold for people touching vehicles at road crossings. Those at the edge of RoW were set to minimize the risk of possible health effects due to long-term exposures.

The existing limits are aimed at the population protection. No fixed rules exist for workers in substations. Such circumstance is due to the fact that, because of station automation and conservative design, the time spent by workers in high electric fields is small.

The rationale for setting limits is generally given in preliminary safety studies, which are requested to regulatory agencies, often as an effect of the pressure of public opinion. As a consequence, the justification of limits does not ever rely on firm scientific data.

Table 11. Typical values of maximum electric field and right-of-way widths in USA (Zaffanella, 1985)

Voltage (kV)	Max. field strength (kV/m)	RoW width (m)
345	5	45
500	8	53
765	10	76

Table 12. Limit values of the electric field strength under power lines recommended in USA (Zaffanella, 1985)

State	Max. field strength (kV/m)	Comments
Minnesota	8	Within RoW
Montana	7	At road crossings
	1	At edge of RoW
New Jersey	3	At edge of RoW
New York	11.8	Within RoW
	11	At private roads
	7	At public roads
	1.6	At edge of RoW
North Dakota	8	Within RoW
Oregon	9	Within RoW

A good example of the effects of public pressure on regulatory agencies is reported by Repacholi (1985) for the State of New Jersey. The resolution of the Commission on Radiation Protection states that, in the absence of evident health effects, it is impossible to propose a scientifically rigorous field exposure threshold, below which undesirable effects will not occur. Nevertheless, concern among the public about possible adverse human health effects makes it advisable to provide some interim guidance with respect to an electric field strength level below which it is very unlikely that adverse effects occur. The limit value of 3 kV/m was chosen since on one hand no health effect has been reported for exposures at or below this level, and on the other one this limit does not constitute at present an economic constraint for power lines constructors in the State.

USSR

Soviet Union has been the first country to issue official exposure standards for electric fields generated by transmission lines, as a consequence of the concern raised by the already mentioned findings of Asanova and Rakov (1966) and of Sazonova (1967). A first regulation document (USSR, 1970) regarding workers employed at ac substations and transmission lines operating at 400, 500, and 750 kV was approved by the USSR Ministry of Health on 29 October 1970, to be enforced in 1971. Five years later, the standard was replaced by a new regulation (USSR, 1975), applying to workers in substations or on transmission lines operating at 400 kV and above, as to include people working on the new EHV systems at 1,150 kV. The regulation was enforced on 1 January, 1977, to last five years. After 1 January, 1981, the standard had to be either changed or reaffirmed, but at present we have no information on any official document issued on this matter.

According to the standard, the duration of exposure to electric fields is limited depending on the field strength, as shown in Table 13.

The rationale for the Soviet standard is given by Lyskov et al. (1975). Studies conducted in USSR since 1962 revealed major effects of high-voltage alternate-current (HVAC) electric fields on exposed workers. These effects were due to direct influence of the electric field, to electric discharges, and to currents leaving the human body.

Regarding to the direct influence of fields, the reaction of the organism is non-specific, it can develop after a relatively long time (2-5 months), it has a long term consequence and pronounced cumulative effects, and is largely dependent on the characteristics of the individual.

Prolonged exposure to high fields can give functional disturbances of the central nervous system and of the heart-vascular system, lower the sexual capability, and induce changes in the blood composition.

Electric discharges, that a man may experience when touching objects in an electric field, can lead to disturbance of stimulation and inhibition reflexes in the outer layer of the brain. In the case of large discharges, heart fibrillation cannot be excluded.

Moreover, if the current intensity through the body exceeds the let-go threshold, the man cannot tear himself away from the charged object and muscle spasms can even lead to his death.

In 1984, also a standard for the protection of the population was issued by the USSR Ministry of Health (USSR, 1984). It regards exposures to fields generated by transmission lines operating at 330 kV and above.

No protection is required on the contrary in the case of lines of 220 kV and below, provided they conform to Soviet norms for the design of high-voltage systems.

Table 13. Electric field exposure limits for workers in installations of 400 kV and above in USSR (1975)

Electric field strength (kV/m)	Permitted exposure time per day (min)
5	Unrestricted
10	180
15	90
20	10
25	5

Notes: 1. If workers are exposed to electric fields of 10 kV/m or more for the full time permitted by the standard, they must remain in fields of 5 kV/m or less for the rest of the day.
2. Workers exposed to 10 kV/m or above can remain for the permitted time, provided they are not subjected to spark discharges.

The standard provides maximum permissible levels of electric field strengths, as listed in Table 14.

To protect the general population from the effects of electric fields, fields, the regulation also establishes the possibility to create a so called "sanitary protection zone", defined as the territory along a transmission line route in which the field strength exceeds 1 kV/m.

The establishment of such a zone is allowed at the moment of the design of a transmission line, if no other mean of reducing the field strength exists. The maximum permissible distances of the boundaries of the sanitary protection zones from the projection onto the ground of the outside phase conductor are listed in Table 15.

The sanitary protection zone is not an exclusion zone: it is rather conceived as an area in which long stays are to be avoided, as well as activities which may lead to electrical discharges or to other unpleasant or dangerous effects. In this zone it is therefore prohibited to handle fuels and inflammable substances, to repair equipment, to operate metallic machineries which are not earthed or screened.

At the time of design, the routes of power lines are to be selected as to leave out of the sanitary protection zone dwelling houses, public buildings, parkings, car service stations, and fuel stores. If that is not reasonably feasible, it is permitted to leave existing houses in the protection zone, provided the field strength inside the buildings and on the open ground is below the limits of Table 14, or it is reduced within these limits by screening systems, such as metal roofs, wire cages, fences, etc.

No dwelling house can anyway be left in the sanitary protection zones of 750 kV and above. After the licensing of the lines, the construction of buildings and facilities within the protection zone is forbidden.

Table 14. Maximum permissible levels of electric field strength for the general public in USSR

Characteristics of the area	Max. E field (kV/m)
Inside residential premises	0.5
Residential built-up areas	1
Rural inhabited areas, suburban areas, health resorts, areas of prospective urban development, etc.	5
Cross of lines with main roads	10
Uninhabited accessible areas, arable land	15
Uninhabited, not readily accessible areas	20

Notes: 1. The specified limits are for unperturbed fields, at a height of 1.8 m from the ground level, or from the floor level in the case of rooms.
2. For field strengths over 1 kV/m measures must be taken to prevent people to be affected by electric discharges and leakage currents.

Table 15. Half-width of the sanitary protection zone in USSR

Operation voltage (kV)	Half-width (m)
330	20
500	30
750	40
1,150	55

Recommendations are also given for the activities which may take place within the sanitary protection zone. For example, it is suggested that arable land be used for growing crops, the raising of which does not involve any manual tasks. If other plants are cultivated, which require metallic wires to help growing, it is recommended that the wires be aligned perpendicular to the transmission line. The existence of the protection zone must be marked at points where the line crosses roads, by signs indicating that the stop of motor vehicles is forbidden.

Special additional protection rules are given for lines operating at very high voltages, namely 750 and 1150 kV. As already mentioned, existing houses must be moved outside the confines of the protection zone. Persons under eighteen are forbidden to perform any working activity in the protection zone.

Finally, besides the definition of the sanitary protection zone, a minimum distance of the line from the boundary of inhabited locations is to be established at the time of design. This distance is of 250 m for 750 kV lines, and of 300 m for 1150 kV lines, and may be reduced only where the route of the line is confined by natural obstacles, or in exceptional circumstances, but in no case can be less than the limits of Table 14. Anyway, if the lines approach rural inhabited locations to within distances less then those indicated above, the conductors must be arranged as to reduce the field strength below 5 kV/m, i.e. the limit set in Table 14 for non-permanent stay.

The basis for the Soviet regulation regarding the general population is essentially the same as for the workers, and is briefly reported in the standard document itself. It is stated that an electric field close to transmission lines can have harmful effects on humans, which increase as the strength of the electric field increases. A distinction is made among the types of effect, namely a direct effect, the effect of electrical discharges from charged bodies, and the effect of leakage currents. The nature and the consequences of the direct effect are not specified, but it is stated that they increase both with the field strength and the duration of the exposure.

The Ministry for Public Health of the USSR (USSR, 1985) has issued the only official standard regulating exposure to 50 Hz magnetic fields. In the standard, a distinction is made between continuous and pulsed field. The time of exposure is limited, depending on the pulse characteristics, as shown in Table 16.

The standard seems to have been developed for arc welding, which constitutes the main source of occupational exposure to pulsed magnetic fields at power frequency.

Table 16. Maximum permissible levels of 50 Hz magnetic fields (A/m) in the USSR (1985)

Duration of exposure (h)	Pulse characteristics		
	Continuous or t>0.02 s; T<2 s	1 s<t<60 s T>2 s	0.02 s<t<1 s T>2 s
1.0	6,000	8,000	10,000
1.5	5,500	7,500	9,500
2.0	4,900	6,900	8,900
2.5	4,500	6,500	8,500
3.0	4,000	6,000	8,000
3.5	3,600	5,600	7,600
4.0	3,200	5,200	7,200
4.5	2,900	4,900	6,900
5.0	2,500	4,500	6,500
5.5	2,300	4,300	6,300
6.0	2,000	4,000	6,000
6.5	1,800	3,800	5,800
7.0	1,600	3,600	5,600
7.5	1,500	3,500	5,500
8.0	1,400	3,400	5,400

t = pulse width duration
T = pulse pause duration

No rationale for this standard appears to have been published so far.

STANDARDS ISSUED BY INTERNATIONAL ORGANIZATIONS

WHO and IRPA

Following the recommendations of the United Nations Conference on the Human Environment held in Stockholm in 1972, and in response to a number of World Health Resolutions, and the recommendation of the Governing Council of the United Nations Environment Programme, a programme on the integrated assessment of the health effects of environmental pollution was initiated in 1973. The programme, known as the WHO Environmental Health Criteria Programme, has been implemented with the support of the Environment Fund of the United Nations Environment Programme. In 1980, the Environmental Health Criteria Programme was incorporated into the International Programme on Chemical Safety (IPCS). The result of the Environmental Health Criteria Programme is a series of criteria documents.

The International Radiation Protection Association (IRPA) initiated activities concerned with non-ionizing radiation by forming a Working Group on Non-Ionizing Radiation in 1974. This Working Group later became the International Non-Ionizing Radiation Committee (INIRC) at the IRPA meeting in Paris in 1977.

The IRPA/INIRC reviews the scientific literature on non-ionizing radiation, makes assessments of the health risks of human exposure to such radiation and, in cooperation with the Environmental Health Division of the World Health Organization, has undertaken responsibility for the development of environmental health criteria documents on non ionizing radiation.

For each kind of radiation the criteria document includes an overview of the physical characteristics, of measurement techniques and instrumentation, of sources and application, and finally of scientific literature on biological and health effects.

These criteria documents become the scientific bases for the development of guidelines on exposure limits and codes of safe practice.

Two joint WHO/IRPA Task Groups on Environmental Health Criteria Documents for Extremely Low Frequency Fields (WHO/IRPA, 1984) and Magnetic Fields (WHO/IRPA, 1987) respectively, reviewed and revised existing scientific literature. They made an evaluation of the health risks of exposure to electromagnetic fields, considered rationales for the development of human exposure limits, and gave advice and recommendations for further research, so that on the basis of more definite data bases a critical revision of the existing standards can take place on an international basis, possibly reaching unique and certain safety criteria.

WHO suggests (WHO/IRPA, 1984) that, whilst it would be prudent in the present state of scientific knowledge not to make unqualified statements about the safety of intermittent exposure to electric fields, there is no need to limit access to regions where the field strength is below about 10 kV/m. Even at this field strength, some individuals may experience uncomfortable secondary physical phenomena such as spark discharge, shocks, or stimulation of the tactile sense.

Furthermore, according to WHO, it is not possible from present knowledge to make a definitive statement about the safety or hazard associated with long-term exposure to sinusoidal electric fields in the range of 1-10 kV/m.

In the absence of specific evidence of particular risk or disease syndromes associated with such exposure, and in view of experimental findings on the biological effects of exposure, it is recommended that efforts be made to limit exposure, particularly for members of the general poulation, to levels as low as can be reasonably achieved.

IRPA/INIRC interim guidelines on limits of exposure to 50/60 Hz electric and magnetic fields, both for workers and the general public, are now in progress. A draft document (IRPA, 1987) has been distributed to IRPA Associate Societies, as well as to a number of competent institutions and individual experts for comments.

The rationale for the IRPA/INIRC guidelines is given in the WHO Environmental Health Criteria Documents for ELF fields (WHO, 1984) and for magnetic fields (WHO, 1987), but so far no definitive document has been issued.

REFERENCES

Asanova, T. P., and Rakov, A. I., 1966, The State of health of persons working in electric field of outdoor 400 and 500 kV switchyards, Translated by G. Knickerbocker, 1975, in: "Study in the USSR of Medical Effects of Electric Power Systems. IEEE Special Publ. no. 10," IEEE Pow. Eng. Soc., Piscataway, N.J.

Bernhardt, J. H., 1984, Electromagnetic fields: potential hazard, Bundesarbeitsblatt, 10:17 (in German).
Bernhardt, J. H., ed., 1986, "Biological Effects of Static and Extremely Low Frequency Magnetic Fields," MMV Medizin Verlag, Munich.
Bernhardt, J. H., Rothe, F. K., Dahme, M., 1983, "Hazard for Persons from Electromagnetic Fields". Report STH 2/83, Institute for Radiation Hygiene, Federal Institute of Health, Reimer-Verlag, Berlin (in German).
Banks, R. S., 1986, Regulations of Overhead Power Transmission Line Electric and Magnetic Fields. Syllabus of the International Utility Symposium "Health Effects of Electric and Magnetic Fields: Research, Communication, Regulation," Toronto, September 16-19, 1986, Ontario Hydro, Toronto
CIGRE, 1986, "Electric and Magnetic Fields from Power Transmission Systems. Results of an International Survey," CIGRE Paper 36-09, International Conference on Large High-Voltage Electric Systems, Paris.
FRG, 1986, "Hazards by Electromagnetic Fields; Protection of Persons in the Frequency Range from 0 Hz to 3000 GHz," Standard DIN VDE 0848 (Draft), German Electrotechnical Commission on DIN and VDE.
FRG, 1986, "Safety by electromagnetic fields; Limits of field strengths for the protection of persons in the frequency range from 0 to 30 kHz," Standard DIN VDE 0848, part 4, German Electrotechnical Commission on DIN and VDE.
Grandolfo, M., Michaelson, S. M., and Rindi, A., eds., 1985, "Biological Effects and Dosimetry of Static and ELF Electromagnetic Fields," Plenum Press, New York and London.
IRPA, 1987, "Guidelines on Limits of Exposure to Electric and Magnetic Fields at Power Frequencies of 50/60 Hz" (Draft).
Japan, 1976, "Electrostatic Induction Caused by Extra-High Voltage Overhead Transmission Lines," Special document for IERE members No. R-7604. English version of original document published in Japanese by Japan IERE Council, 1975.
Johnson, J., Connelly, K., and Smith, D., 1976, "Influence of Electric Field Effects on 500 kV System Design," CIGRE Paper 31-05, International Conference on Large High-Voltage Electric Systems, Paris.
Lyskov, Y. I., Emma, Yu., S., and Stolyarov, M. D., 1975, The factors of electrical field that have an influence on a human, in: "Three Russian Papers on EHV/UHV Transmission Line and Substation Design," Uhl, Hall and Rich, Division of Charles T. Main, Inc., Boston.
Kabrhel, I., 1985, Answers to the CIGRE Questionnaire "Survey of the Situation Concerning the Effects of Electric and Magnetic Fields," Doc. 36-85 (SC) 28 IWD.
NRPB, 1982, "Health Protection of Workers and Members of the General Public against the Dangers of Extra Low Frequency, Radiofrequency, and Microwave Radiations: A Consultative Proposal," National Radiological Protection Board, Chilton, Didcot, Oxon.
NRPB, 1986, "Advice on the Protection of Workers and Members of the Public from the Possible Hazards of Electric and Magnetic Fields with Frequency Below 300 GHz: A Consultative Document," National Radiological Protection Board, Chilton, Didcot, Oxon.
Pilatowicz, A., 1985, Answers to the CIGRE Questionnaire "Survey of the Situation Concerning the Effects of Electric and Magnetic Fields", Doc. 36-85 (SC) 09 IWD.
Poland, 1980, "Decree of 5 November 1980 of the Cabinet Concerning Particular Rules of Protection against Non-Ionizing Electromagnetic Fields Dangerous for People and the Environment," Official Gazette of the Polish People's Republic No. 25. Warsaw, 17 November.
Repacholi, M. H., 1985, Standards on static and ELF electric and magnetic fields and their scientific basis, in: "Biological Effects and

Dosimetry of Static and ELF Electromagnetic Fields," M. Grandolfo, S. M. Michaelson, and A. Rindi, eds., Plenum Press, New York and London.

Sazonova, T. E., 1967, A physiological assessment of work conditions in 400-500 kV open switching yards. Translated by G. Knickerbocker, 1975, in: "Study in the USSR of Medical Effects of Electric Power Systems. IEEE Special Publ. no. 10," IEEE Pow. Eng. Soc., Piscataway, N.J.

Shah, K. R., 1979, "Review of State/Federal Environmental Regulations Pertaining to the Electrical Effects of Overhead Transmission Lines: 1978," U.S. Dept. of Energy Publication HCP/EV-1802, DOE, Washington.

Silney, J., 1985, Effects of low-frequency, high intensity magnetic fields on the organism, in: "Proc. Int. Conf. on Electric and Magnetic Fields in Medicine and Biology," Inst. Elect. Eng., London.

Smith, D. C., 1985, Answers to the CIGRE Questionnaire "Survey of the Situation Concerning the Effects of Electric and Magnetic Fields", Doc. 36-85 (SC) 18 IWD.

Tenforde, T. S., and Kaune, W. T., 1987, Interaction of extremely low frequency electric and magnetic fields with humans, Health Physics, 53:585.

UNIPEDE, 1985, "Rapport d'Activite' du Group d'Etudes Medicales," Athenes, 9-14 juin 1985, Rapport 85.f.90.1 (in French).

USSR, 1970, Rules and regulations on labor protection at 400, 500, and 750 kV A.C. substations and overhead lines of industrial frequency," USSR Ministry of Health, Moscow. Translated by G. Knickerbocker, 1975, in: "Study in the USSR of Medical Effects of Electric Power Systems. IEEE Special Publ. no. 10," IEEE Pow. Eng. Soc., Piscataway, N.J.

USSR, 1975, "Occupational Safety Standards System. Electrical Fields of Current Industrial Frequency of 400 kV and Above. General Safety Requirements," Standard no. 12.1.002-75, National Standards Committee, Moscow (in Russian).

USSR, 1984, "Sanitary Standards and Regulations for Protecting the Population from the Effects of an Electric Field Produced by Industrial-Frequency A.C. Overhead Transmission Lines," USSR Ministry of Health, Moscow (In Russian. Translation available from the Central Electricity Generation Board, London, as Document CE 8264).

USSR, 1985, "Maximum Permissible Levels of Magnetic Fields with the Frequency 50 Hz," Document 3206-85, USSR Ministry of Health, Moscow (in Russian).

WHO, 1981, "Environmental Health Criteria 16. Radiofrequency and Microwaves," World Health Organization, Geneva.

WHO, 1984, "Environmental Health Criteria 35. Extremely Low Frequency (ELF) Fields," World Health Organization, Geneva.

WHO, 1987, "Environmental Health Criteria 69. Magnetic Fields," World Health Organization, Geneva.

Yasui, M. 1985, Answers to the CIGRE Questionnaire "Survey of the Situation Concerning the Effects of Electric and Magnetic Fields," Doc. 36-85 (SC) 09 IWD.

Zaffanella, L. E., 1985, Answers to the CIGRE Questionnaire "Survey of the Situation Concerning the Effects of Electric and Magnetic Fields," Doc. 36-85 (SC) 21 IWD.

INSTRUMENTATION FOR ELECTROMAGNETIC FIELDS EXPOSURE EVALUATION AND ITS ACCURACY

Santi Tofani* and Motohisa Kanda**

* Laboratorio di Sanità Pubblica – Sezione Fisica, 10015 Ivrea, Italy and Istituto Nazionale di Fisica Nucleare – Sezione di Torino, 10125 Torino, Italy

** National Institute of Standards and Technology, Electromagnetic Fields Division Boulder, Co 80303 – U.S.A.

INTRODUCTION

Setting the exposure limits to RF and microwave (MW) electromagnetic fields, contained in the safety standards, required a great effort to the scientific community which spent many years working to evaluate biological effects and to accomplish a dosimetry.

A great part of the above mentioned effort could be invalid if, simultaneously, we did not work to minimize the errors related to the evaluation of the exposure to such fields in the frame of the procedures for physical surveillance (Bernardi et al., 1985).

We stress the fact that measurements in the field of physical surveillance may give errors in the evaluation of the exposure level of several dBs unless periodic calibration is implemented.

Purposes of this chapter are the following:

a) Analysis of the field strength meter instruments characteristics;

b) Analysis of the exposition methods, with particular attention to the methods used in calibration activities and to their accuracies;

c) Analysis of the errors due to the field strength meters utilization in the routine of physical surveillance, with special attention to near-field evaluation errors. These strongly depend on measurement procedures.

FIELD STRENGTH METER INSTRUMENTS CHARACTERISTICS

The field strength meters used to assess the potential RF and MW hazard are generally made up of three basic parts: 1- the antenna, 2- the detector, and 3- the metering unit. The antenna and the detector generally constitute the probe (sensor). In Fig. 1 the scheme of a classic field meter is illustrated.

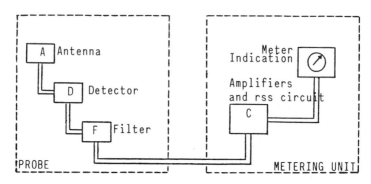

Fig. 1. Scheme of a field strength meter.

The antenna generally consists of an element that responds only to the E field (short dipole) or only to the H field (small loop). A new kind of probe has recently been developed to measure the electric and magnetic fields simultaneously so that an electromagnetic environment may be characterized more completely, particularly in near-field measurements (Kanda, 1984). Recently field strength meters with the antenna consisting of a passive sensing element like the electro-optic modulator have been used. Electro-optic techniques in making field meter instruments are improving (Ma and Kanda, 1986). To make the probe isotropic (that is its response is indipendent of the direction of propagation of electromagnetic field) the antenna is built with three elements (dipoles or loops) arranged in a mutually orthogonal way as shown in Figs. 2 and 3.

The detector produces a continuous current or voltage proportional to the square of the RF/MW current or voltage amplitude, respectively. The more commonly used detectors are semiconductor diodes and thermocouples. The filter network, located between the detector and the metering unit, will locally block the RF currents that may be an important source of measurement errors. The filter network rectifies RF waveforms at the detector and transmits only the "DC" component to the amplifiers.

Non-linear circuitry included in the metering unit can be used to give the rss (root sum of square) values of the three orthogonal field components detected by the three dipoles or loops. In this way the meter indication will be proportional to the rms value of the total field (E or H), that is:

$$E = (E_x^2 + E_y^2 + E_z^2)^{1/2} \text{ or } H = (H_x^2 + H_y^2 + H_z^2)^{1/2}$$

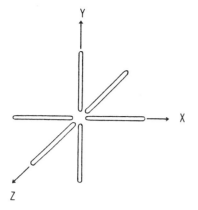 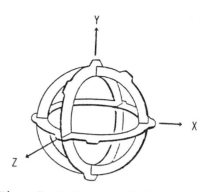

Fig. 2. Orthogonal dipoles of an electric field probe.

Fig. 3. Orthogonal loops of a magnetic field probe.

To make meaningful measurements the field strength meter should have the following main characteristics (ANSI, 1987):

1) The maximum dimension of the antenna should be less than the minimum wavelength of the radiation being measured;

2) The probe must respond to a particular electromagnetic field parameter and not have spurious responses;

3) The probe should be isotropic within 1 dB;

4) The field meter instrument should employ a self-contained power supply, isolated from external fields by appropriate shielding and filter decoupling;

5) The instrument should indicate the following parameters:
   a) the electric field strength in volts per meter (V/m) or the mean squared electric field strength in square volts per square meter (V^2/m^2),
   b) the magnetic field strength in amperes per meter (A/m) or the mean squared magnetic field strength in square amperes per square meter (A^2/m^2),
   c) average "equivalent plane wave" power density in watts per square meter (W/m²) or in milliwatts per square centimeter (mW/cm²). When this kind of instrument is used in the near-field, the results of the measurement must be transformed in terms of electric or magnetic field strength depending on the kind of antenna used;

6) To facilitate spatial and time averaging and to protect the operator from hazardous field during the measurement procedures, the field meter instrument should be equipped with a recorder output and/or other means (for example a repeater connected to the meter via fiber optic);

7) The response of the instruments should be fairly flat independently of the frequency of radiation and its dynamic range should range at least from -10 dB (10%) to 5 dB (300%) of the exposure limits valid in the country where the instrument is used;

8) It is desirable that the instruments are provided with the following special functions:
   a) memorizing of the maximum value accurring during the measurement period,
   b) a data logging function which can provide a real time average of the measured fields utilizing an averaging time specified by the user;

9) The instrument should be provided with total calibration uncertainties no greater than ± 2 dB, or better yet (but difficult to achieve) within ± 1 dB.

EXPOSURE AND CALIBRATION METHODS

A reliable calibration of the instruments used for electromagnetic field level measurements is essential to ensure compliance with safety standards and to allow comparison of the exposure level measurements, for example in bioelectromagnetic activities performed in different laboratories and countries. Existing calibration methods are based on the assumption that knowledge of the standard electromagnetic field level can be obtained by means of measurements, calculation or by a combination of both. For calibration, the instrument is exposed to the standard electromagnetic field and the reading value is compared with the value of the standard field, which is known by means of measurements and/or calculations.

The main methods for the generation of standard fields, suitable for instruments calibration, can be grouped as follows:

A) the free-space standard field method;
B) guided-wave method;
C) transfer standard method.

The exposure methods listed above, in particular A) and B) methods, are commonly used also in bioelectromagnetic experimental researches and in electromagnetic compatibility activities.

For the purpose of giving more complete information, we also give a brief description of the reverberating chambers characteristics. The use of these chambers in bioelectromagnetics and in electromagnetic compatibility is a relatively new one.

Free space standard method

This method is generally adopted to generate microwave electromagnetic fields ($f \geq 300$ MHz) by means of radiators such as horns, open-ended guides, or other directive antennas whose gains are known with great accuracy. The antennas are located inside shielded rooms to prevent external contributions to the field and lined with anechoic material (pyramidal polyurethane foam impregnated with conductive carbon and/or magnetic iron particles) to reduce the electromagnetic energy reflected from the chamber walls to a predictable low level. The shielding material is generally composed of rigid laminated modular panels where two thin sheets of steel (0.5 mm thick) are laminated to 20 mm solid core made of plywood, thus forming a sandwich resulting in a rigid structural panel. This material gives the possibility of an attenuation of about 100 dB over the frequency range from 1 kHz to 20 GHz and beyond. The anechoic material ability of preventing reflections depends basically on the pyramid height, the frequency, and the radiation incidence angle; the height must not be shorter than one half wavelength.

As an example Fig. 4 shows, for a widely adopted kind of anechoic material (heigth of 66 cm), the obtainable reflection coefficient versus frequency and angle of incidence. The angle of incidence represents the angle between the incident radiation and the normal to the surface lined by the anechoic material. Thus an angle of incidence of 0° means that the incident radiation is orthogonal to the surface covered by the pyramids while an angle of incidence of 90° means that the incident radiation is parallel to the wall covered by the pyramids.

We can see that for radiation incident perpendicular to the plane containing the base of the pyramidal elements (incident angle of 0° degrees) and for frequencies f≥300 MHz, a good reflection coefficient (≤-30 dB, that is 0.1%) can be reached. To obtain the same reflectivity at lower frequencies it would be necessary to use taller elements that, besides being more expensive, would diminish the useful space.

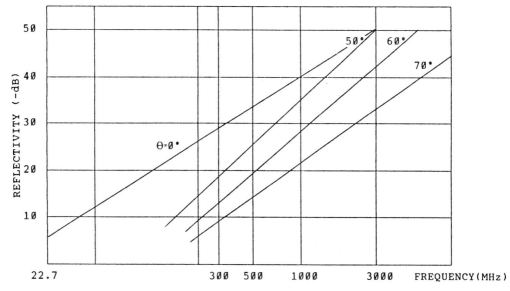

Fig. 4. Electromagnetic field reflectivity versus frequency and angle of incidence θ (data obtained from pyramidal absorbers with height of 66 cm).

Fig. 4 shows that the reflectivity, that is proportional to the ratio between the radiation intensity reflected from the surface and that incident to the surface itself, decreases when the frequency increases because the wavelength of the radiation becomes smaller compared with the height of the pyramids. Moreover, and irrespective of the frequency, the reflectivity increases as the angle of incidence radiation increases.

The electromagnetic field radiated by the antenna can be generated and controlled by means of an instrumental chain similar to that showed in Fig. 5. The signal is produced by a function generator, amplified, and sent to the directive antenna through a bidirectional coupler. This device allows the withdrawal of two signals proportional to the incident and reflected power respectively. These signals are sent to two bolometric sensors connected to a power meter.

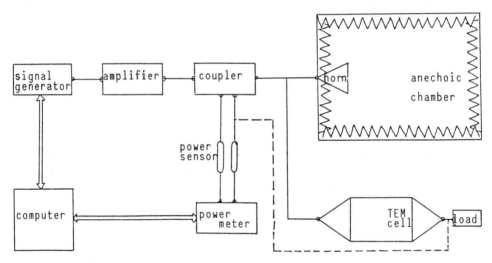

Fig. 5. Instrumental chain used to produce and control electromagnetic fields.

The power meter connected to the two bolometric sensors represents the most important part of the chain. In fact, from the reading of the incident and reflected power, we can know the power available to the antenna, and thus the electromagnetic field strength (power density and electric and magnetic field strengths) used to calibrate the field strength meter. A computer interfaced with the signal generator and with the power meter allows a continuous monitoring and controlling of the frequency and of the available power to the antenna or other radiating systems like the TEM cell.

The electromagnetic field power density in the chamber, S, is given by the following equation:

$$S = \frac{P_a \, G \, R}{4 \pi r^2} \quad [W/m^2] \quad (1)$$

where:
- P_a is the power (in watts) available to the antenna,
- G is the antenna gain, that is defined as $4\pi r^2$ times the ratio of the maximum power density radiated in a prescribed direction to the imput power radiated by the antenna itself;
- R is a factor which accounts for the chamber influence, hence this parameter is quantified from the reflectivity discussed before; and
- r is the distance in meters from the antenna aperture.

If measurements are taken in far-field conditions, we obtain the rms electric (E) and magnetic (H) field strengths by means of the following equations:

$$E = (S\ 377)^{1/2} \quad [V/m] \quad (2)$$

$$H = (S\ /\ 377)^{1/2} \quad [A/m] \quad (3)$$

where:
377 is the free space impedance expressed in ohms.

Pyramidal horns or open-ended waveguides (OEG) are used as transmitting antennas, positioned in the access doorway with their apertures inside the plane defined by the absorber tip points on the chamber wall. The net power delivered to the transmitting antenna is the difference between incident and reflected powers as measured with a dual directional coupler (four-ports).

Electromagnetic field measurements in an anechoic chamber are usually performed in the near-field region of a transmitting standard antenna. The magnitude of the field is computed from a theoretically derived expression. As an example, the antennas now used at the U.S. National Institute of Standards and Technology (NIST) consist of a series of open-ended waveguides at frequencies below 450 MHz and a series of rectangular pyramidal horns above 450 MHz.

At present there is not complete, rigorous theory of OEG near-field gain. As suggested by Yaghjian (Yaghjian, 1982) the near-field of an open-ended, unflanged, rectangular waveguide is calculated from near-field power patterns, which are determined from theoretically predicted far-field power patterns by use of the plane-wave scattering theorem (Kerns, 1981). Far-field power patterns are predicted by inserting the E and H fields of the propagating TE_{10} mode into the Stratton-Chu formula and integrating over the aperture of the OEG (Stratton, 1941).

An assumption of a TE_{10} mode field in the waveguide is only an approximation because it neglects the higher order, evanescent modes caused by the blunt edge of the OEG. The accuracy of the near-field reconstructed using estimated far-field data will depend on the accuracy of the far-field estimation.

The classical result of the aperture integration using TE_{10} mode field in the Stratton-Chu formulas gives an E-plane pattern that agrees well with the measured pattern (Risser, 1949), while the H-plane pattern does not agree as well. The agreement of the H-plane may be improved using a Yaghjian correction term, which accounts for the contribution from the fringe currents (Yaghjian, 1983). A comparison between theoretical and experimental data on the radiating near-field of an OEG is presented in a recent paper (Wu and Kanda, to be published).

The approach used at NIST to establish a standard (i.e., calculable) field at frequencies above 450 MHz involves the use of a series of rectangular pyramidal horns. In deriving the near-field gain of a pyramidal horn by the Kirchhoff method, Schelkunoff accounted for the effect of the horn flare by introducing a quadratic phase error in the dominant mode field along the aperture coordinates. Geometrical optics and single diffraction by the aperture edges yield essentially the Kirchhoff results. The proximity effect in the Fresnel zone can also be approximated by a quadratic phase error in the aperture field. Taking into account the preceding considerations, the near-zone gain (G) of a pyramidal horns is given by (Kanda et al.,1986):

$$G = \frac{32ab}{\pi \lambda^2} R_E R_H \qquad (4)$$

where
R_E and R_H are gain reduction factors due to the E-plane and H-plane flares, respectively; they depend upon frequency, horn dimensions, and distance to the on-axis field point (Kanda and Orr, 1987).

a and b are the width (H-plane flare) and height (E-plane flare) of the rectangular horn aperture, and
λ is the free-space wavelength.

Guided wave method

In guided wave methods, the electromagnetic fields are generated in cavities such as a waveguide cell or a TEM cell (Crawford, 1974). These facilities are generally employed for frequencies lower than 300 MHz because at these frequencies the use of anechoic chamber, besides being quite expensive,would introduce reflection errors. The above mentioned facilities are completely shielded (in this way there is no problem for workers' and instruments' exposure), very stable, and quite cheap.

In this frequency range, rectangular-waveguide transmission cells can be used as calibrating chambers. Electromagnetic power is transmitted through the guide to a matched resistive load, and the upper frequency in each guide is limited to that in which operation is in the dominant TE_{10} mode. The E and H fields at the guide center can be obtained (Kanda et al, 1986) from the following equations:

$$E_y = (2ZP / ab)^{1/2} \qquad (5)$$

$$H_x = (2P / Zab)^{1/2} \qquad (6)$$

where
E_y is the electric field strenght (V/m),

H_x is the electric field strength (A/m),
P is the total power flow in the guide, (W),
a,b are the cross section dimensions of the guide, (m),
and
Z is the wave impedance (Ω) in the cell $=377/[1-(\lambda/2a)^2]^{1/2}$

The TEM cell can be seen as an expanded transmission line operating in the transverse electromagnetic (TEM) mode, hence the name. As shown in the scheme reported in Fig. 6 the main body of the cell consists of a rectangular outer conductor and a flat center conductor located midway between the top and bottom walls. The dimensions of the main section and the tapered ends of the cell are chosen to provide a 50-ohm impedance along the entire length of the cell.

In the center of the cell, halfway between the center conductor and the top (or bottom) wall, the E-field is vertically polarized and quite uniform. Also, the wave impedance (E/H) is very close to the free space value of 377 ohms. The main disadvantage related to their utilization is the small amount of useful space for the instrument calibration; as a matter of fact, this space is only a few percent of the space occupied by the cell.

The rms electric field strength inside the cell is:

$$E = (P_a \ Z_o/d)^{1/2} \quad [V/m] \quad (7)$$

where:
P_a is the available power to the cell input,
Z_o is the characteristic impedance of the cell expressed in ohms, and
d is the distance from the septum or center conductor to the wall, expressed in meters.

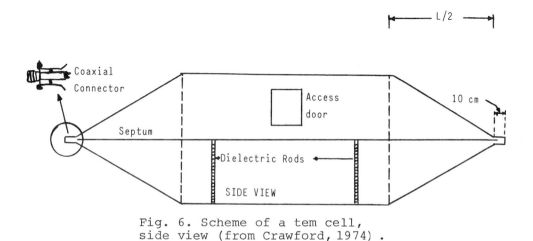

Fig. 6. Scheme of a tem cell, side view (from Crawford, 1974).

In a TEM cell the electromagnetic field is very nearly that of a plane wave, hence the rms magnetic field strength is:

$$H = \frac{E}{377} \quad [A/m] \quad (8)$$

The wave travelling through the cell has essentially a free-space impedance (377 Ω), thus providing a close approximation to a far-field plane wave propagating in free space.

The upper useful frequency for a cell is limited by distortion in the test field due to multimoding and resonances that occur within the cell at frequencies above the cell's multimode cutoff. Resonant frequencies, f_{res}, associated with these modes can be found from the expression (Kanda et al, 1986):

$$f_{res} = [\, f^2_{mn} + (\, vl/2L\,)^2\,]^{1/2} \quad (9)$$

where:

f_{mn} are the frequencies of the higher order mode(s) excited in the cell;
v is the wave phase velocity (= 3×10^8 m/s);
L is the resonant length of the cell in meters;

and:

l, m and n are integers corresponding to multiples of the resonant length and the particular waveguide mode.

Transfer standard method

This method is based on a precise calibration, carried out with one of the above reported methods, of a stable and reliable probe which is then used as a transfer standard. The standard probe is used for field strength measurement in a fixed point of a field which is produced by arbitrary sources and is unknown. Subsequently in that fixed point is placed the probe to be calibrated and its reading is compared with that of the standard probe.

As we will see later, the accuracy of this method is worse than the previous reported ones. As a matter of fact this method is a relative calibration, not an absolute one.

Reverberating chamber

A reverberating chamber is a shielded room with a rotating stirrer to mix the field generated by a transmitting antenna. This chamber is large in terms of wavelength so that many modes can be excited at the same time. In addition, the moving reflecting surfaces give a time varying field pattern. The resulting field within a region inside the chamber can therefore be described in terms of an isotropic plane wave spectrum with random polarization and with a uniformly homogeneous time average intensity. In addition, the reverberating chambers are able to provide a very efficient conversion of source power to high-strength fields inside a shielded enclosure (Corona et al., 1976).

The above reported characteristics are very useful in bioelectromagnetic experimental researches; for example they may offer a brilliant solution to the problem of supplying to a small moving laboratory animal an absorbed power constant over the exposure period independently of both position and orientation of the subject (d'Ambrosio et al.,1980). On the other hand the isotropic plane wave spectrum with random polarization may be a limiting factor in characterizing the object under test in electromagnetic compatibility and calibration activities.

The use of reverberating chambers is good for very high frequency application (commonly used frequencies range from a few hundred MHz to 20 GHz and above) and may serve as a powerful supplementary tool to TEM cells.

ACCURACY IN THE CALIBRATION OF FIELD STRENGTH METERS

The evaluation of the field strength meters calibration errors is related to the accuracy levels with which the standard electromagnetic fields are generated. The evaluation of these accuracy levels is generally carried out assuming the most pessimistic hypothesis of the error propagation. This method used for the accuracy evaluation is called the worst case method (w.c.m.). It is based on the assumption that each error source acts in the same direction with its maximum amplitude. Denoting with Q a quantity that is function of magnitudes X_k, that is:

$$Q = f(X_1, \ldots, X_k), \tag{10}$$

the error σ_Q expressed in dB (for more information about error evaluation in the w.c.m. see Appendix) and calculated with the w.c.m. is:

$$\sigma_Q = \sum_{1}^{m} {}_k \sigma_k \tag{11}$$

Applying w.c.m. to equations (1),(2),(3),(7) and (8) we have the overall error relative the electromagnetic field generated respectively in the anechoic chamber (σ_E, σ_H) and in the TEM cell (σ_E1, σ_H1) respectively:

$$\sigma_E = \sigma_H = \sigma_{Pa} + \sigma_G + \sigma_R \tag{12}$$

$$\sigma_E1 = \sigma_H1 = \sigma_{Pa} + \sigma_{Zo} + \sigma_d + \sigma_D \tag{13}$$

where the symbol σ denotes the error expressed in dB and the subscripts denote the related parameter. The error σ_r is not reported in equation (12) since it is negligible. The error σ_D of the equation (13) accounts for the field non-homogeneity in the TEM cell due to the presence of the hazard-probe under test.

The relative parameters σ_i which may determine the total error in a typical anechoic chamber facility and in a TEM cell are reported (Tofani et al., 1986) in Tables 1 and 2 respectively.

Table 1. Typical errors for anechoic chamber facilities.

FREQ. (MHz)	PARTIAL ERRORS (dB)			TOTAL ERRORS (dB)
	σ_{Pa}	σ_G	σ_R	$\sigma_E = \sigma_H$
250	0.64	0.25	0.15	1.04
300	0.62	0.25	0.15	1.02
400	0.64	0.26	0.15	1.06
500	0.66	0.32	0.15	1.14
600	0.62	0.25	0.15	1.02
700	0.61	0.24	0.15	1.00
800	0.67	0.26	0.15	1.08
900	0.65	0.25	0.15	1.06
990	0.67	0.28	0.15	1.06

In both cases the total error relative to the electric and magnetic field strengths is about 1 dB (~12%) independent of the frequency.

Tables 1 and 2 show that the biggest partial error (near to 0.6 dB) is that related to the available power (Pa) because this parameter includes different measurements, incident power and reflected power respectively. Furthermore, in Table 2 we can see that the error σ_D, which accounts for the field non-homogeneity in the TEM cell, is 0.25 dB (~3%).

Table 2. Typical errors for TEM cell facilities

FREQ. (MHz)	PARTIAL ERRORS (dB)				TOTAL ERRORS (dB)
	σ_{Pa}	σ_d	σ_{Zo}	σ_D	$\sigma_E = \sigma_H$
0.1	0.65	0.01	0.17	0.25	1.08
1	0.65	0.01	0.17	0.25	1.08
50	0.58	0.01	0.17	0.25	1.01
100	0.66	0.01	0.17	0.25	1.09
150	0.62	0.01	0.17	0.25	1.05
200	0.61	0.01	0.17	0.25	1.04
250	0.60	0.01	0.17	0.25	1.03

With regard to the transfer standard method, the mean error obtainable can oscillate, in the best case, around 2 dB. The main source of errors, for this kind of calibration, is due to the difference in the receiving pattern of the two probes.

International intercomparison activities are periodically

carried out among between Laboratories specialized in the generation of electromagnetic standard fields in order to minimize the errors on the generation and measurement of the fields themselves.

Recently the "Laboratorio di Sanità Pubblica" (LSP) of Ivrea, Italy, was charged with the study and development of electromagnetic standard fields measurements by the Italian "Istituto Nazionale di Fisica Nucleare" (INFN) as reference for a number of its research Centers, and the U.S. National Institute of Standards and Technology (NIST) concluded a first intercomparison in the frequency range 0.5-500 MHz. For the intercomparison a commercial field strength meter has been used as travelling standard.

The calibration was performed at an electric field strength of 50 V/m and at the frequencies of 0.5, 1, 10, 100, 300 and 500 MHz. Results are reported in Fig. 7 where the calibration factor C (defined as $C = E_{espected} / E_{measured}$, expressed in dB) is shown versus the frequency for the two probe orientations we adopted with respect to the E field. The difference between the calibration factors obtained in the two Laboratories is decidedly lower than the theoretically evaluated uncertainty with which the electromagnetic fields are known by both Laboratories ($\sim \pm 1$ dB). The results of the two Laboratories differ on the average of about 0.3 dB with a dispersion of 0.2 dB.

These results suggest that the uncertainties with which electromagnetic fields are actually generated are lower than what theoretically expected, under the assumption of w.c.m. As a matter of fact, if we consider that the electromagnetic fields have been generated and controlled in the two Laboratories using different systems, it is unlikely that the same systematic error in both systems in the evaluation of the calibration factor C reported in Fig. 7.Therefore we can consider the two sets of a data are independent each from the other, and the probability that the results of two independent distributions of data referring to the same magnitude (E field) differ more than the errors with which the are known is almost zero.

Therefore the effective error in the knowledge of the Electric field strength in both Laboratories is probably closer to 0.3 ± 0.2 dB than 1 dB; on the other hand the r.s.s. method applied to the partial errors of Tables 1 and 2 gives a total error near to 0.7 dB instead of about 1 dB obtained using the w.c.m.

MAIN SOURCES OF ERRORS IN THE PRACTICAL USE OF FIELD STRENGTH METERS

We have analyzed some of the more generally adopted methods of field meters calibration and their related accuracies. Unfortunately it is not possible to reach the same accuracy when these field meters are used for environmental measurements. Some of the principal reasons for this are the following.

1) The field meters are generally calibrated in conditions

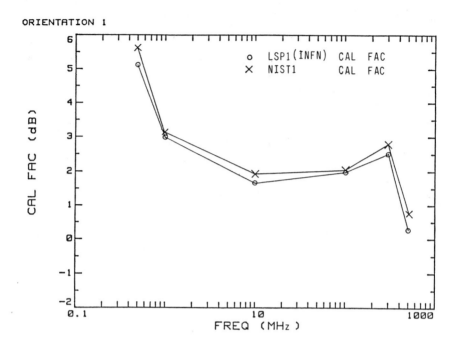
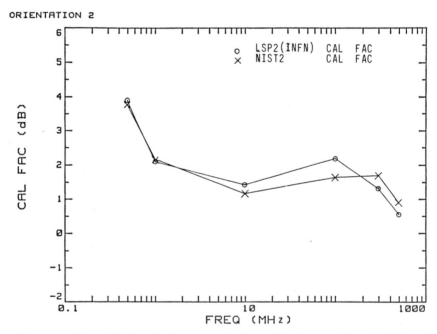

Fig. 7. Results of the intercomparison between the NIST and LSP-Ivrea. Data referring to two different probe orientations.

of far-field. These conditions can rarely be found in practice and the field meters can supply very different values from the "true values" when used in non plane-wave conditions.

2) In most cases during the field meter calibration only the probe is exposed to the electromagnetic field, while in the common practice the whole system (meter, cable, etc.) is exposed. Non negligible errors can result from spurious responses of such components to the electromagnetic field.

3) The field meter response is generally not uniform depending on probe orientation with respect to the incident field (response anisotropy). The response difference due to such anisotropy can vary from 1 to 2 dB.

Hence, the careful utilization of such meters for the physical surveillance can allow measurement accuracies of 2 to 3 dB in most cases. However, in the case of near-field measurements, measurements at very high frequencies, and the measurements in zones where reflecting structures are present, this extra error can be so high that special procedures are needed to minimize the errors.

For example, in the case of the surveillance near the plastic sealers where the exposure is characterized by:

a) temporal non-uniformity of the emission from these devices;
b) spatial non uniformity of the operator exposure conditions;
c) operator mobility,

we need to have a continuous monitoring of these parameters to obtain an accurate estimate both of the exposure and of the $SAR_{NEAR-FIELD}$ (Chatterjee et al., 1982) Without the adoption of rigorous measurement procedures based upon a continuous monitoring and recording of the electromagnetic field strength, the evaluation of the occupational exposure may be affected by errors up to 5 dB (Tofani and Agnesod, 1987).

APPENDIX

If x is a given magnitude with an absolute error s_x, the error σ_x of x expressed in dB is:

$$\sigma_x = 10 \lg [(x + s_x)/x] \quad (1)$$

Denoting with Q a quantity that is a function of magnitudes x_k that is:

$$Q = f(x_1, \ldots, x_k) \quad (2)$$

the error σ_Q expressed in dB and calculated with the w.c.m. is:

$$\sigma_Q = \sum_1^m \sigma_k \quad (3)$$

From the equations (1) and (3) we can obtain the relative error (s_Q/Q) (calculated with the w.c.m. but not expressed in dB) just removing the logaritmic operator in the equation (3). The result is:

$$(s_Q/Q) = [\prod_{1}^{m}{}_k (x_k + s_k)/ x_k] - 1 \qquad (4)$$

This relative error can be compared with the one found by using a more realistic method of combining errors, that is gaining in popularity, the root sum of the squares (r.s.s.) method. The r.s.s. method is based on the fact that most of the measurement errors although systematic and not random are independent of each other. From the Gauss law of error propagation, the above relative error calculated with this method is:

$$(s_Q/Q) = [\sum_{1}^{m}{}_k (s_k/x_k)^2]^{1/2} \qquad (5)$$

As expected, from the comparison between equations (4) and (5), the following inequality is always satisfied:

$$\prod_{1}^{m}{}_k [(x_k + s_k)/x_k] - 1 > [\sum_{1}^{m}{}_k (s_k/x_k)^2]^{1/2} \qquad (6)$$

To prove the validity of the above inequality we can suppose that all the m relative errors (s_k/x_k) are equal. With this hypothesis the inequality (6) becomes:

$$[1+(s_k/x_k)]^m > 1 + m^{1/2} (s_k/x_k) \qquad (7)$$

From the property of the exponential function we know that $(1+\alpha)^n > 1+n\alpha$ with $\alpha > 0$ and n integer; therefore the inequality (6) is demonstrated. The error evaluation using the w.c.m. is then the maximum one; another reason of its use in experimental electromagnetic activities is the fact that all the measurement results are generally expressed in dB. It becomes therefore easy to assess the total error only by summing individual errors.

REFERENCES

ANSI, 1987, American National Standard Recommended Practice for the measurement of potentially hazardous electromagnetic fields - RF and microwave, ANSI Draft, May 11, 1987 (Ed. Petersen R. C.)

Bernardi, P., Boggio, M., Checcucci, A., Grandolfo, M., Righi, E., Tamburello, C., 1985, Criteri in tema di sorveglianza fisica per il rischio da campi elettromagnetici a radiofrequenze e microonde, L'Elettrotecnica, LXXII (2): 135.

Chatterjee, I., Gandhi, O. P., and Hagmann, M. J., 1982, Numerical and experimental results for near-field electromagnetic absorption in man, IEEE Trans. Microwave Theory Tech., 30:2000.

Corona, P., Latmiral, G., Paolini, E., and Piccioli, L., 1976,

Use of a reverberating enclosure for measurements of radiated power in the microwave range, *IEEE Trans. Electromagn. Compat.*, EMC-18:54.

Crawford, M. L., 1974, Generation of standard EM fields using TEM transmission cells, - *IEEE Trans. Electromagn. Compat.*, EMC-16:189.

d'Ambrosio, G., Di Meglio, F., and Ferrara, G., 1980, Multimode time varying enclosures for exposure and dosimetry in bioelectromagnetic experiments, *Alta Frequenza*, XLIX:89.

Kanda, M., 1984, An electromagnetic near-field sensor for simultaneous electric- and magnetic-field measurements, *IEEE Trans. Electromagn Compat.*, EMC-26: 102.

Kanda, M., Larsen, E. B., Borsero, M., Galliano, P. G., Yokoshima, I., and Nahman, N. S., 1986, Standards for electromagnetic fields measurements, - *Proc. IEEE*, 24: 120.

Kanda, M., and Orr, R. D., 1987, Near-field gain of a horn and an open-ended waveguide: comparison between theory and experiment, *IEEE Trans. Antennas Propagat.*, AP-35:33.

Kerns, D. M., 1981, Plane-wave scattering-matrix theory of antennas and antenna-antenna interactions, *Nat. Bur. Stand. Monogr. 162*.

Ma, M. T., and Kanda, M., 1986, Electromagnetic compatibility and interference metrology, *NBS Tech. Note 1099*, June 1986, Chap.7, (Availability: superintendent of documents, U.S. Government Printing Office Washington, D.C. 20402).

Risser, J. R., 1949, Waveguide and horn feeds, in "Microwave Antenna Theory and Design", S. Silver, ed., McGraw Hill, New York.

Stratton, J. A., 1941, "Electromagnetic Theory", McGraw Hill, New York.

Tofani, S., Anglesio, L., Agnesod, G., and Ossola, P., 1986, Electromagnetic standard fields: generation and accuracy levels from 100 KHz to 990 MHz, *IEEE Trans. Microwave Theory Tech.*, 34:832.

Tofani, S., and Agnesod, G., 1987, Dosimetry of occupational exposure to RF radiation: measurements and methods, *IEEE Trans. Microwave Theory and Tech.*, 35:594.

Wu, D. I., and Kanda, M., to be published, Theoretical and experimental data comparison of the radiating near-field of an open-ended rectangular waveguide, to be published in *IEEE Trans. Antennas Propagat.*

Yaghjian, A. D., 1982, Efficient computation of antenna coupling and fields within the near-field region, *IEEE Trans. Antennas Propagat.*, AP-30:113.

Yaghjian, A. D., 1983, Approximate formulas for the far-fields and gain of open-ended rectangular waveguide, *Nat. Bur. Stand. NBSIR 83-1689*.

NUMERICAL METHODS

Om P. Gandhi
Electrical Engineering Department
University of Utah
Salt Lake City, Utah, 84112, U. S. A

INTRODUCTION

Knowledge of the SAR and temperature distributions in the human body in response to electromagnetic exposures is of basic interest in the assessment of biological effects and medical applications of electromagnetic energy. Several methods have been described in the literature for numerical calculations of SAR distributions (Taflove and Brodwin, 1975b; Chen and Guru, 1977; Hagmann et al., 1979). While experiments must be done to verify the accuracy of any theoretical procedure, the numerical methods do allow a detailed modeling of the anatomically-relevant inhomogeneities for a human that are not easy to model experimentally. A couple of excellent review articles summarize the status of the field (Durney, 1980; Spiegel, 1984), though these are somewhat outdated at the present time. Maxwell's equations have been solved in both the integral and differential equation form for the distribution of electric and magnetic fields and mass-normalized rates of energy deposition (specific absorption rates or SARs). For temperature calculations the bioheat equation is solved for a cylindrical thermal model where the human body is represented by six cylindrical segments for the head, torso, arms, hands, legs and feet (Stolwijk, 1970; Stolwijk and Hardy, 1977 and Stolwijk, 1980). Each segment consists of four concentric cylinders representing layers of skin, fat, muscle and core tissue. Recognizing that such formulations are incapable of properly accounting for the painstakingly-obtained SAR distributions, we have developed an inhomogeneous thermal block model of man using heterogeneous thermal properties obtained from the anatomic data for 476 cells (Chatterjee and Gandhi, 1983). Since SARs are now available with a higher degree of resolution than ever in the past, a 5432-cell inhomogeneous thermal model of the human body has also been developed recently (Gandhi and Hoque - to be published).

Because of the limitations on the length of the paper, we will focus on the numerical methods for SAR calculations and not discuss the thermal models for temperature calculations. To simplify computational requirements, several authors have also used two-dimensional formulations for individual cross sections of the body. Since this is not an accurate description of the three-dimensional (3-D) nature of the human body problem, we do not discuss these techniques in this paper. The focus instead is on techniques that lend themselves to high resolution 3-D modeling of the body for electromagnetic exposures either far-field or near-field in nature. For SAR calculations, the most promising techniques at the present time are: finite-difference time-domain (FDTD) method (Spiegel, 1984, 1985; Sullivan et al., 1987, 1988) and the sinc-basis fast Fourier transform (FFT) - conjugate gradient (CG) method (Borup et al. - to be published) which is a modified version of the traditional method of moments (Chen and Guru, 1977; Hagmann et al., 1979). For lower frequencies where quasi-static approximations may be made (≤ 40 MHz for the human body), the impedance method has been found to be highly efficient numerically and has been used for a number of applications (Armitage et al., 1983; Gandhi et al., 1984; Gandhi and DeFord, 1988; Orcutt and Gandhi, 1988).

METHOD OF MOMENTS

The method of moments (MOM - Harrington, 1968) has been extensively used to calculate SAR distributions in block model representations of the human body (Chen and Guru, 1977; Hagmann et al., 1979; Livesay and Chen, 1974; Gandhi et al., 1979; Hagmann and Gandhi, 1979). In this procedure the internal fields $E(r)$ within the body are solved from the solution of the electric field integral equation (EFIE) obtained from the Maxwell's equations. The EFIE for an isotropic, nonmagnetic, arbitrarily inhomogeneous body is as follows:

$$E^i(r) = E(r) - (k_o^2 + \nabla\nabla \bullet) \int_{v'} (\varepsilon_r^*(r) - 1) \, E(r') \, g \, (r - r') \, dv' \tag{1}$$

where $E^i(r)$ are the incident fields at the individual locations within the body, if the body were absent; $E(r')$ are the unknown total fields at the various locations, k_o is the wave number in free space and ε_r^* are the complex dielectric constants at the various locations

$$g \, (r - r') = \frac{e^{-jk_o|r-r'|}}{4\pi|r-r'|}$$

is the 3-D free-space Green's function and the integral v' is taken over the entire volume of the body.

It should be noted that the unknown fields $E(r)$ are involved both outside and inside the integral sign of Eq. (1). With pulse-basis functions, which implies that the electric field is assumed constant over the volume of the individual cells, the EFIE can be expressed in the form of a matrix equation

$$\overleftrightarrow{A} \bullet E = E^i \tag{2}$$

where $\overleftrightarrow{A}$ is a 3N x 3N matrix that describes coupling between the various cells, E is a vector of 3N unknown field components, 3 for each of the N cells into which the body is divided and E^i is a vector of 3N known field components that would be incident at the same N-cells if the body did not exist. Solution of this matrix equation by conventional methods such as Gauss elimination or the conjugate gradient method with matrix multiplication requires order $(3N)^2$ storage and order $(3N)^2 - (3N)^3$ computation steps. Both of these computational requirements expand rapidly as the number of cells N is increased. Consequently, it has not been possible to apply the traditional MOM to block model representation of the human body with N larger than 1132 (DeFord et al., 1983). In spite of the difficulty of handling a larger number of cells, image theory has been used to modify the Green's function in lieu of the free-space Green's function allowing thereby SAR calculations in models of man with grounding and reflector effects (Hagmann and Gandhi, 1979).

FFT - CG METHOD OF MOMENTS

Recognizing the convolutional nature of the EFIE (Eq. (1)) iterative algorithms based on the highly efficient fast Fourier transform (FFT) have been developed which have recently allowed the heterogeneous modeling of the human body by up to 5607-cell models (Borup et al. - to be published). In this method as with all moment methods, the numerical solution of Eq. (1) begins with approximation of the source polarization current $J_p = (\varepsilon_r^* - 1) \, E$ with a finite basis expansion. In the present approach the tensor product sinc basis is used

$$J(x,y,z) = \frac{\delta^3}{\pi^3}\sum_n\sum_m\sum_l J(n\delta, m\delta, l\delta)\frac{\sin\frac{\pi}{\delta}(x-n\delta)\sin\frac{\pi}{\delta}(y-m\delta)\sin\frac{\pi}{\delta}(z-l\delta)}{(x-n\delta)(y-m\delta)(z-l\delta)} \quad (3)$$

where δ is the grid increment. Inserting Eq. (3) into Eq. (1) and enforcing the equality of Eq. (1) at the grid points results in a linear system of equations.

$$\mathbf{E}^i(i,j,k) = \mathbf{E}(i,j,k) + \sum_n\sum_m\sum_l \overline{\overline{G}}(i-n, j-m, k-l) \cdot \left[\left(\varepsilon^*_{nml}-1\right)\mathbf{E}(n,m,l)\right]$$

where

$$\overline{\overline{G}}(n,m,l) = (k_o^2 + \nabla\nabla\cdot)\left\{\frac{e^{jk_o r}}{4\pi r} * \delta^3 \frac{\sin(\frac{\pi x}{\delta})}{\pi^3} \frac{\sin(\frac{\pi y}{\delta})}{x} \frac{\sin(\frac{\pi z}{\delta})}{y} \right\}\Bigg|_{\substack{x=n\delta\\y=m\delta\\z=l\delta}} \quad (4)$$

and * denotes a 3-D convolution.

Because an equally-spaced grid has been used, Eq. (4) has inherited the convolution form of Eq. (1). This form can be exploited to provide a means of computing the sum in Eq. (4) in order $N\log_2 N$ arithmetic operations. In addition, no matrix need be stored since only the 3-D FFTs of the components of $\overline{\overline{G}}$ are needed. This provides an efficient method to compute the matrix products $A\overline{x}$ and $A^H\overline{x}$ where A is the matrix operator implicit in Eq. (4) (the superscript H denotes the complex conjugate transpose matrix). In order to solve the system (Eq. (4)), the efficient calculation of the matrix-vector products is used in the implementation of a conjugate gradient-type iteration method (CGM). The CGM has recently become very popular as a means of solving linear systems obtained with the MOM. The most-used version of this algorithm was originated by Hestenes and Stiefel (1952) and introduced by Sarkar et al. (1981) for electromagnetic problems. Unfortunately, this method often requires a large number of iterations to obtain a converged solution. This is because the CGM applies only to positive definite, hermition matrices. This method is applied to nonhermition systems by first "squaring" the original system, $A\overline{x} = \overline{y}$ with the conjugate transpose matrix to give the normal equations, $A^H A\overline{x} = A^H \overline{y}$. The conjugate gradient method is then applied to the hermition system $A^H A$. The problem with this is that the formation of the normal equations squares the spectral condition number of the linear system. Because the number of iterations needed for convergence is asymptotically equal to the square root of the condition number, this procedure can lead to slow convergence.

To alleviate this difficulty we have replaced the CGM with a biconjugate gradient method (BCGM) developed by Fletcher (1975). The BCGM applies directly to nonhermition matrices without the formation of the normal equations and has been found to converge nearly as rapidly as the CGM applied to similarly-sized and conditioned hermition systems. Our experience is that the use of the BCGM results in a drastic reduction in the number of iterations needed over the CGM applied to the normal equations for systems obtained from Eq. (4).

After testing the accuracy of the new formulation with homogeneous and layered lossy dielectric sphere test problems, we have used it with 5607-cell man models (2.62 cm cubical cells) exposed to 30 and 100 MHz plane waves with and without a ground plane. Typical run times are on the order of 10 minutes using CRAY II.

THE FINITE-DIFFERENCE TIME-DOMAIN (FDTD) METHOD

Perhaps the most successful and the most promising of the numerical methods for SAR calculations at the present time is the finite-difference time-domain (FDTD) method. This method was first proposed by Yee (1966) and later developed by Taflove and colleagues (1975 a,b; 1980); Umashankar and Taflove (1982), Holland (1977) and Kunz and Lee (1978). Recently it has been extended for calculations of the distribution of electromagnetic fields in a man model for incident plane-waves (Spiegel, 1984, 1985; Sullivan et al., 1987, 1988; Chen and Gandhi, 1989b) and for exposures in the near fields (Wang and Gandhi, 1989; Chen and Gandhi, 1989a). In this method, the time-dependent Maxwell's curl equations

$$\nabla \times \mathbf{E} = -\mu \frac{\partial \mathbf{H}}{\partial t} \quad , \quad \nabla \times \mathbf{H} = \sigma \mathbf{E} + \varepsilon \frac{\partial \mathbf{E}}{\partial t} \quad (5)$$

are implemented for a lattice of cubic cells. The components of **E** and **H** are positioned about a unit cell of the lattice (see Fig. 1) and evaluated alternately with half-time steps. The goal of the method is to model the propagation of an electromagnetic wave into a volume of space containing dielectric and/or conducting structures by time stepping, i.e., repeatedly implementing a finite-difference analog of the curl equations at each cell of the corresponding space lattice. The incident wave is tracked as it first propagates to the dielectric structure and interacts with it via surface current excitation, spreading, penetration and diffraction. Wave tracking is completed when a sinusoidal steady-state behavior for **E** and **H** fields is observed for each lattice cell. A second approach is to illuminate the body with a pulse of radiation in the time-domain. If the medium is nondispersive (i.e., its dielectric properties do not change with frequency), this method has the advantage of giving the performance of the medium at various frequencies from the inverse Fourier transform of the calculated fields. Though Spiegel and his collaborators (1984, 1985) have followed this approach, we have used the sinusoidal variation of the illuminating fields since the dielectric properties of the tissues are highly dependent on frequency.

The Maxwell's equations (5) can be rewritten as difference equations. We assume the media to be nonmagnetic, i.e., $\mu = \mu_0$ and a grid point of the space defined as (i, j, k) with coordinates (iδ, jδ, kδ) where δ is the cell size.

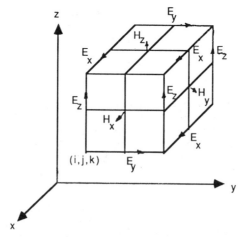

Fig. 1. A unit cell of the Yee lattice indicating the positions for the various field components.

Upon substituting

$$\tilde{E} = \sqrt{\frac{\varepsilon_o}{\mu_o}} \frac{E}{2} \qquad (6)$$

the equations in difference form for E_z and H_z for examples, can be written as:

$$\tilde{E}_z^{n+1}(i,j,k+1/2) = CA_z(i,j,k+1/2)\tilde{E}_z^n(i,j,k+1/2)$$

$$+ CB_z(i,j,k+1/2) \cdot \begin{bmatrix} H_y^{n+1/2}(i+1/2, j,k+1/2) - H_y^{n+1/2}(i-1/2,j,k+1/2) \\ + H_x^{n+1/2}(i,j-1/2, k+1/2) - H_x^{n+1/2}(i,j+1/2, k+1/2) \end{bmatrix} \qquad (7)$$

$$H_z^{n+1/2}(i+1/2, j+1/2, k) = H_z^{n-1/2}(i+1/2, j+1/2, k)$$

$$+ \begin{bmatrix} \tilde{E}_x^n(i+1/2, j+1, k) - \tilde{E}_x^n(i+1/2, j, k) \\ + \tilde{E}_y^n(i, j+1/2, k) - \tilde{E}_y^n(i+1, j+1/2, k) \end{bmatrix} \qquad (8)$$

where

$$CA_z(i,j,k+1/2) = \left[1 - \frac{\sigma_z(i,j,k+1/2)\delta t}{2\varepsilon_z(i,j,k+1/2)}\right]\left[1 + \frac{\sigma_z(i,j,k+1/2)\delta t}{2\varepsilon_z(i,j,k+1/2)}\right]^{-1} \qquad (9)$$

$$CB_z(i,j,k+1/2) = \frac{\varepsilon_0}{4}\left[\varepsilon_z(i,j,k+1/2) + \frac{\sigma_z(i,j,k+1/2)\delta t}{2}\right]^{-1} \qquad (10)$$

where the superscript n denotes the time $n\,\delta t$ in terms of the incremental time or time step δt. The time step δt is determined by the cell size and must satisfy the stability condition (Taflove and Brodwin, 1975a)

$$v_{max}\,\delta t < \left[\frac{1}{\delta x^2} + \frac{1}{\delta y^2} + \frac{1}{\delta z^2}\right]^{-1/2} \qquad (11)$$

where v_{max} is the maximum wave phase velocity within the model. For the cubic cell model $\delta x = \delta y = \delta z = \delta$, the following relationship is usually taken

$$\delta t = \frac{\delta}{2C_{max}} \qquad (12)$$

where C_{max} is the maximum velocity of electromagnetic waves in the interaction space. We have taken C_{max} corresponding to velocity of electromagnetic waves in air. The ε and σ may be different for E_x, E_y, and E_z in the inhomogeneous media, because, as shown in

Fig.1 the components of **E** are not positioned at the same point for the same cell in Yee's algorithm. Thus we have $\varepsilon_x, \varepsilon_y, \varepsilon_z, \sigma_x, \sigma_y$, and σ_z, for each of the cells.

A basic problem with any finite-difference solution of Maxwell's equations is the treatment of the field components at the lattice truncation. Because of the limited computer storage, the lattice must be restricted in size. Proper truncation of the lattice requires that any outgoing wave disappear at the lattice boundary without reflection during the continuous time-stepping of the algorithm. An absorption boundary condition for each field component is therefore needed at the edge of the lattice. In our formulation, the absorption boundary conditions in second approximation derived by Mur (1981) are used. The condition for E_z at i = 1 for example can be written as:

$$\tilde{E}_z^{n+1}(1,j,k+1/2) = -\tilde{E}_z^{n-1}(2,j,k+1/2)$$

$$+ \frac{C_\ell \, \delta t - \delta}{C_\ell \, \delta t + \delta}\left[\tilde{E}_z^{n+1}(2,j,k+1/2) + \tilde{E}_z^{n-1}(1,j,k+1/2)\right]$$

$$+ 2\left(\frac{\delta}{C_\ell \, \delta t + \delta} - \frac{(C_\ell \, \delta t)^2}{\delta(C_\ell \, \delta t + \delta)}\right)\left(\tilde{E}_z^n(1,j,k+1/2) + \tilde{E}_z^n(2,j,k+1/2)\right)$$

$$+ \frac{(C_\ell \, \delta t)^2}{2\delta(C_\ell \, \delta t + \delta)}\begin{bmatrix}\tilde{E}_z^n(1,j+1,k+1/2) + \tilde{E}_z^n(1,j-1,k+1/2) + \tilde{E}_z^n(2,j+1,k+1/2) \\ + \tilde{E}_z^n(2,j-1,k+1/2) + \tilde{E}_z^n(1,j,k+3/2) + \tilde{E}_z^n(1,j,k-1/2) \\ + \tilde{E}_z^n(2,j,k+3/2) + \tilde{E}_z^n(2,j,k-1/2)\end{bmatrix}$$

(13)

where C_ℓ is the phase velocity of the medium at the lattice truncation.

AN ANATOMICALLY-BASED INHOMOGENEOUS MODEL OF MAN

As previously described in Sullivan et al. (1987,1988), the model for the human body is developed from information in the book, "A Cross-Section Anatomy," by Eycleshymer and Schoemaker (1911). This book contains cross-sectional diagrams of the human body which were obtained by making cross-sectional cuts of spacing of about one inch in human cadavers. The process for creating the data base of the man model was the following: first of all, a quarter-inch grid was taken for each single cross-sectional diagram and each cell on the grid was assigned a number corresponding to one of the 14 tissue types given in Table I, or air. Thus the data associated with a particular layer consisted of three numbers for each square cell: x and y positions relative to some anatomical reference point in this layer, usually the center of the spinal cord, and an integer indicating which tissue that cell contained. Since the cross-sectional diagrams available in Eycleshymer and Schoemaker (1911) are for somewhat variable separations, typically 2.3-2.7 cm, a new set of equispaced layers were defined at 1/4-inch intervals by interpolating the data onto these layers. Because the cell size of quarter-inch is too small for the memory space of present-day computers, the proportion of each tissue type was calculated next for somewhat larger cells of size half-inch or one inch combining the data for 2x2x2 = 8 or 4x4x4 = 64 cells of the smaller dimension. Without changes in the anatomy, this process allows some variability in the height and weight of the body. We have taken the final cell size of 1.31 or 2.62 cm (rather than half-inch or one inch) to obtain the whole-body weight of 69.6 kg for the model. The number of cells either totally or partially within the human body for the two models are 41,256 and 5628, respectively.

Table 1. Tissue properties used at various frequencies.

Tissue Type	Mass density 1000 kg/m³	27.12		100 MHz		350 MHz	
		σ S/m	ε_r	σ S/m	ε_r	σ S/m	ε_r
Air	0.0012	0.0	1.0	0.0	1.0	0.0	1.0
Muscle	1.05	0.74	106	1.0	74	1.33	53.0
Fat, bones	1.20	0.04	29	0.07	7.5	0.072	5.7
Blood	1.00	0.28	102	1.1	74	1.2	65.0
Intestine	1.00	0.29	60	0.55	36	0.66	26.5
Cartilage	1.00	0.04	29	0.07	7.5	0.072	5.7
Liver	1.03	0.51	132	0.62	77	0.82	50.0
Kidney	1.02	0.79	209	1.0	90	1.16	53.0
Pancreas	1.03	0.69	206	1.0	90	1.16	53.0
Spleen	1.03	0.69	206	0.82	100	0.9	90.0
Lung	0.33	0.17	34	0.34	75	1.1	35.0
Heart	1.03	0.64	210	0.75	76	1.0	56.0
Nerve, brain	1.05	0.45	155	0.53	52	0.65	60.0
Skin	1.00	0.74	106	0.55	25	0.44	17.6
Eye	1.00	0.45	155	1.9	85	1.9	80.0

THE IMPEDANCE METHOD

For low frequency dosimetry problems where quasistatic approximation may be made (≤ 40 MHz for the human body), the impedance method has been found to be highly efficient numerically (Armitage et al., 1983; Gandhi et al., 1984, 1988; Orcutt and Gandhi, 1988). In this method, the biological body or the exposed part thereof is represented by a three-dimensional network of impedances whose individual values are obtained from the complex conductivities $\sigma + j\omega\varepsilon$ of the various regions of the body. Using homogeneous and layered cylindrical bodies (DeFord, 1985) or spherical bodies (Orcutt and Gandhi, 1988) as test cases, where the analytic solutions are available (Harrington, 1961), we have established the maximum frequency limit for the impedance method to be on the order of 30-40 MHz for a tissue-equivalent cylinder or sphere of diameter 30 cm, similar to the dimensions of the human torso. In order to establish the frequency limit, solutions were obtained for cases of TE plane-wave irradiation of the cylindrical or the spherical body using the impedance method and compared with the analytical solutions at various frequencies. The effect of the incident electric field is to introduce currents into the model given by $j\omega\varepsilon_0 \mathbf{E}_{ext}$, where $\mathbf{E}_{ext}$ is the external electric field (incident + scattered) at the surface of the body. The contribution from the H-field is to set up emfs in the circuit loops of the model given from Faraday's law of induction by $j\omega\mu_o H_n \delta^2$ where H_n is the magnetic field normal to the plane of the loop of area δ^2. Using the anatomically-based models of man (nominal one-half or one-inch resolution), the impedance method has to date been used for the following applications:

1. **Calculation of SAR distributions for operator exposure to RF induction heater.** In the article by Gandhi and DeFord (1988), SARs are obtained for an anatomically-based model of the human torso for spatially-varying vector magnetic fields because of a 450 kHz RF induction heater.

2. **SAR distributions in the anatomically-based model of the human body for linear or circularly-polarized spatially-variable magnetic fields representative of magnetic-resonance imagers.** In a recent paper by Orcutt and Gandhi (1988), a 1.31 cm resolution model of the human body is used to calculate SAR distributions for RF magnetic fields used for magnetic-resonance imaging of the torso and the head regions of the body. The layer-averaged SAR distribution calculated for a circularly-polarized RF magnetic field (in the cross-sectional plane of the body) is shown in Fig. 2. The layer-numbering system used for the model of the human body is shown in Fig. 3. Each layer is 1.31 cm from its neighbors. Layer 1 is 0.655 cm below the top of the head.

3. **SAR distributions in the human body because of the magnetic field component of incident plane waves.** The calculated layer-averaged SAR distribution for an incident RF magnetic field of 1 A/m at 30 MHz is given in Fig. 6 of the paper on Advances in RF Dosimetry (Gandhi - this issue). It is concluded that the body-averaged SAR because of the magnetic field (0.03 W/kg) is considerably smaller than that for the electric field component of the plane waves at 30 MHz.

4. **SAR distributions because of the spatially-variable vector magnetic fields of magnetrode® applicators for hyperthermia.** Cylindrical current-carrying conductor or magnetrode- type of applicators have been used for hyper-

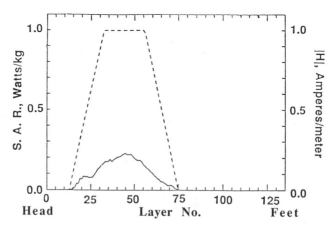

Fig. 2. Layer-averaged SARs for the model of the human body when it is exposed to a magnetic field at 63 MHz, circularly polarized in the x - y plane. The spatial variation of the field strength is shown by the dashed line. The peak SAR for this case, 0.98 Watts/kg, occurred in layer 48. The whole body averaged SAR was 0.03 Watts/kg.

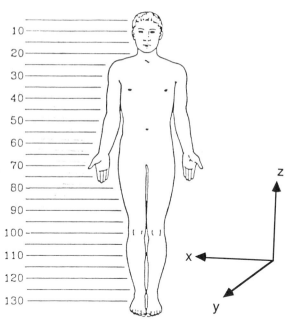

Fig. 3. The layer numbering system used in the model of the human body. Each layer is 1.31 cm from its neighbors. Layer 1 is 0.655 cm from the top of the head.

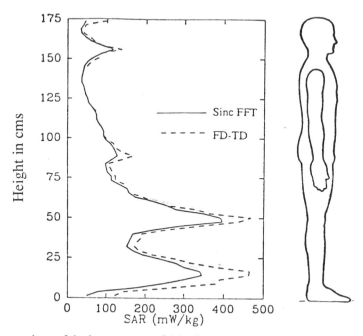

Fig. 4. Comparison of the layer-average SARs for the sinc-function FFT MOM and FDTD methods for an ungrounded man exposed to a vertically-polarized, frontally incident plane wave with intensity 1 mW/cm^2. The calculations were made with the 2.62 cm resolution, 5628-cell model of man. The FFT MOM average SAR is 101 mW/kg. The FDTD method gives 116 mW/kg.

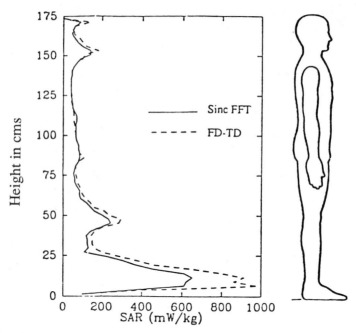

Fig. 5. Comparison of the layer-average SARs for the sinc-function FFT MOM and FDTD methods for an ungrounded man exposed to a vertically-polarized, frontally incident plane wave with intensity 1 mW/cm^2. The calculations were made with the 2.62 cm resolution, 5628-cell model of man. The FFT MOM average SAR is 93 mW/kg. The FDTD method gives 106 mW/kg.

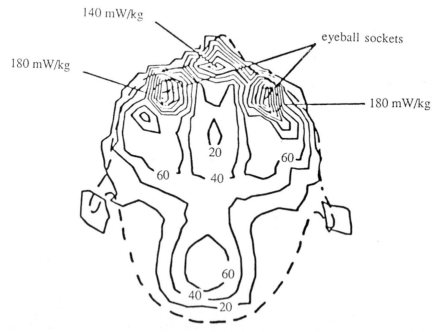

Fig. 6. Contour diagram of the SAR distribution through the head of an inhomogeneous man model. Incident plane wave was 1 mW/cm^2 at 350 MHz. Each contour is 20 mW/kg.

thermia that is caused by induced currents in the body because of time-varying magnetic fields (Elliott et al., 1982). Using the impedance method, we have obtained internal current and SAR distributions accounting for the spatially-variable vector magnetic fields of such an applicator. Whereas a 2-D sectional model of the human body was used earlier (Gandhi et al., 1984) a 3-D model of the human torso has been used by DeFord (1985).

5. SAR distributions because of capacitor-type electrodes used for hyperthermia. Whereas a 3-D, 2.54 cm resolution model of the human torso has been used by DeFord (1985), a 1.31 cm resolution model of the whole human body has recently been used for these applicators (Orcutt and Gandhi, 1989) that are fairly popular in Japan.

6. SAR distributions for interstitial RF needle applicators in irregularly-shaped tumors. In the article by Zhu and Gandhi (1988) a model based on the CT scans of the human head is used to find the optimum locations of the 0.5 MHz RF needles for minimum standard deviations of the SAR distributions in the volume of the identified brain tumor. The procedure described in this article could be used for treatment planning of thermotherapy for a number of tumors for which local hyperthermia by RF needle applicators has been found useful. These tumors include vaginal cuff, colon-rectum, head and neck, etc. (Manning and Gerner, 1981).

Even though isotropic electrical properties have been used for the applications of the impedance method to date, it should be recognized that directional averaging of the electrical properties done at the stage of going from smaller to larger cells would have yielded anisotropic properties for the cells. Furthermore, at lower frequencies the tissue properties are known to be anisotropic which would also result in directionally-dependent impedances which must be used for each of the cells. A natural application of the impedance method would be for dosimetry for 50/60 Hz electromagnetic fields because of power transmission lines.

APPLICATIONS OF THE FDTD AND THE SINC-FUNCTION FFT MOM

We have alluded to several problems where the above approaches may be used for SAR calculations. In the following we describe some of these applications.

1. SAR and induced current distributions for exposures to plane-wave fields for grounded and ungrounded conditions. Both the FDTD and the sinc-function FFT MOM have been used for calculations of SAR distributions in models of man for grounded as well as for ungrounded conditions. As examples, the calculated layer-averaged SAR distributions calculated at 100 MHz are shown in Figs. 4 and 5 for ungrounded and grounded model conditions, respectively. The dielectric properties for the various tissues are taken from Table I. It is interesting to note that even though radically different procedures are used for FFT MOM and the FDTD methods (integral vs. differential equation formulation of the Maxwell's equations), fairly similar layer-averaged SAR distributions are obtained (Figs. 4 and 5). The slight discrepancies between the SARs for the knee and ankle sections may be due to the fact that only few cells have been used for these two sections and the SARs calculated by both methods are somewhat lower than those estimated from measurements of currents induced in human subjects (Gandhi et al., 1986). Higher resolution models are needed for calculations at higher frequencies. The 1.31 cm cell size, 41,256 cell model of man has been used therefore for calculations of SAR distributions at 350 MHz (Sullivan et al., 1988). The contour diagrams of the SAR distribution for the cross section through the eyes is shown in Fig. 6 to illustrate the kind of details that can be obtained. As expected the highest SARs are obtained for the frontal region of the cross section. The calculated whole-body-averaged SARs for grounded and isolated man models are shown in Fig. 7 for the frequency band 20-100 MHz (Chen and Gandhi, 1989b). Also shown for comparison are the experimental data by Hill (1984) with human subjects, and by Guy et al. (1984) with a homogeneous model. Whereas excellent agreement is seen with Hill's data (1984) for human subjects, higher SARs are obtained for the inhomogeneous model used for the present calculations than those for the homogeneous model (Guy et al. 1984). As will be seen later (Figs. 8 and 9) substantial radio-frequency currents are induced in the legs. Because of the large bone content of the legs, an

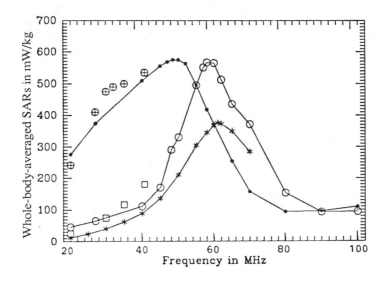

• Grounded inhomogeneous   ⊕ Hill (1984) grounded   ∗ Isolated, Homogeneous
                                                       model (Guy, et al., 1984)
o Isolated, inhomogeneous   □ Hill (1984), ungrounded

Fig. 7. Whole-body-averaged SARs for a grounded and isolated man model. Also shown for comparison are the human subject data of Hill (1984) and the homogeneous, isolated model data of Guy, et al. (1984).

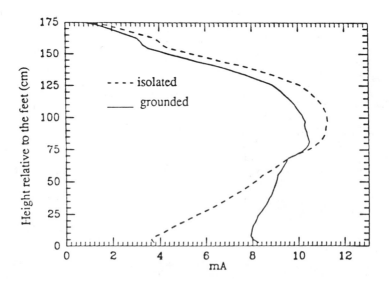

Fig. 8. Induced RF current distributions for a grounded and isolated man model at 50 MHz under 1 V/m plane-wave exposure condition.

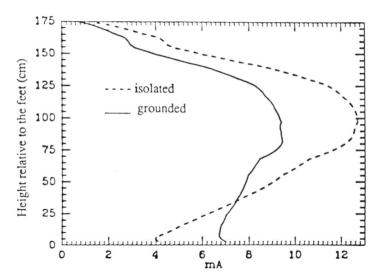

Fig. 9. Induced RF current distributions for a grounded and isolated man model at 60 MHz under 1 V/m plane-wave exposure condition.

inhomogeneous model properly accounts for the lower effective conductivities for the region, which therefore results in higher SARs.

The calculated E-fields can be used to determine the local current densities from the relationship $\mathbf{J} = (\sigma + j\omega\varepsilon)\mathbf{E}$. The z-directed currents for any of the layers in Fig. 3 can be obtained by summing up the terms due to the individual cells in a given layer as follows:

$$I = \delta^2 \sum_i (\sigma_i + j\omega\varepsilon_i) E_{zi} \qquad (14)$$

where δ^2 is the cross-sectional area for each of the cells. The layer-averaged induced current distributions are given in Chen and Gandhi (1989b) for isolated and for grounded conditions for frequencies of 27-90 MHz. The induced current distributions are shown in Figs. 8 and 9 for frequencies of 50 and 60 MHz which were chosen for illustration since they are close to the resonant frequencies for grounded and ungrounded conditions, respectively (see Fig. 7). Since E-fields of 61.4 V/m have been recommended as safe in the ANSI C95.1-1982 RF safety guideline (1982) for the frequency band 30-300 MHz, substantial RF currents are obviously implied from the data given in Figs. 8 and 9.

2. SARs and induced current distributions for leakage fields of a parallel-plate dielectric heater. In a recent article by Chen and Gandhi (1989a), the FDTD method has been used to calculate the local, layer-averaged and whole-body-averaged SARs and induced RF currents in a 5628-cell (cell size = 2.62 cm) model of man for spatially-variable electromagnetic fields of a parallel-plate applicator representative of RF dielectric heaters used in industry. Included in the calculations are the shape and dimensions of the applicator plates (57.6 x 15.7 cm parallel plates with a separation of 2.62 cm) as well as a typical separation of 21 cm to the human operator (geometry shown in Fig. 10). The calculated leakage field components without the man model are in agreement with the experimentally-measured values. The conditions of exposure of the man model considered are: isolated from ground, feet in contact with ground and an additional grounded top plate 13.1 cm above the head to simulate screen rooms that are occasionally used for RF sealers. Also considered is the model with a separation layer of rubber ($\varepsilon_r =$

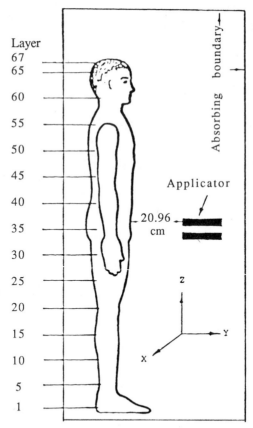

Fig. 10. Geometry of the applicator and man model.

4.2) of thickness 2.62 cm between feet and ground to simulate shoe-wearing condition. The layer-averaged SAR and current distributions calculated for the various conditions are shown in Figs. 11 and 12, respectively, for an irradiation frequency of 27.12 MHz. Since foot currents can be measured for a human subject using a bilayer sensor detailed in Gandhi, et al. (1986), we have made these measurements for a human subject in front of a parallel-plate applicator of dimensions identical to those used for the calculations. The ground plane underneath the current sensor was provided by a 1/16-inch thick aluminum plate of dimensions 1.2 m x 2.4 m. The foot currents have been measured for grounded conditions and for a subject wearing electrical "safety shoes" (size 11, Vibram Manufacturing Company, rubber sole thickness = 1.75 cm). For the so-called isolated condition, the foot current was measured by using a sufficient styrofoam thickness (typically larger than 12.7 cm) under the feet to "isolate" the subject from ground. For such thicknesses the currents were found to be independent of the thickness of styrofoam. The measured currents are shown for comparison also in Fig. 12. The results are in reasonable agreement with the calculated values of the currents through the feet for the various irradiation conditions. A spatially-maximum E-field of 1 V/m rms has been used for the plots shown in Figs. 11 and 12. Since peak E-fields as high as 1000-2700 V/m have been measured at industrial locations typically occupied by the operator, significant internal RF currents on the order of 0.5-2.3 A are projected, scaling the calculated values given in Fig. 12.

3. SAR distributions for annular-phased arrays of aperture and dipole antennas for hyperthermia. One of the challenging tasks in electromagnetic hyperthermia is to develop applicators for selectively heating a deeply-seated tumor in the human body. Because of the difficulty of experimentation with inhomogeneous, let alone

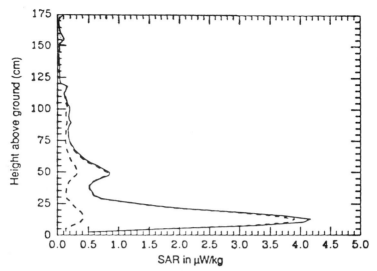

Fig. 11. Layer-averaged SAR distributions for an isolated, grounded and ground-topped man model at 27.12 MHz under near-field exposure conditions. The maximum Erms = 1 V/m at 21 cm in front of the parallel plate applicator. Dashed line (left): isolated man model; dashed line (right): grounded man model; continuous line: ground-topped man model.

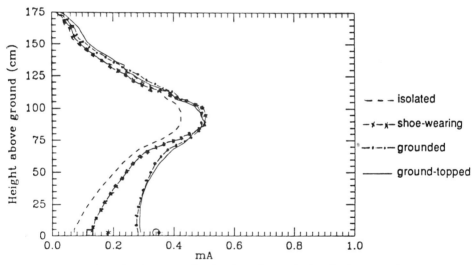

Fig. 12. Induced vertical current distributions for an isolated, shoe-wearing, grounded and ground-topped man model at 27.12 MHz under near-field exposure conditions. The maximum Erms = 1 V/m at 21 cm in front of the parallel plate applicator. Measured foot currents for a human subject: isolated (□), rubber-soled shoes (*), barefoot grounded (O) and ground-topped (●).

anatomically-realistic models, all of the EM applicators at the present time are characterized with homogeneous phantoms. This characterization is relatively worthless and not pertinent to the EM deposition patterns for the human body. The knowledge of the SAR distribution in anatomically-based models is likely to aid in the design of the applicators in regard to the frequency, size, and magnitude and phase of the various elements (if a multielement applicator is used) to obtain the highest SAR in the tumor vis à vis the surrounding tissues.

One of the most popular means of heating internal tissues is the annular-phased arrays (Turner, 1984 a, b). We have used the FDTD method to calculate the SAR distributions in a 17,363-cell, 1.31 cm resolution model of the human torso (Fig. 13) that was surrounded by deionized water to simulate the water bolus that is typically used (Wang and Gandhi, 1989). Eight element arrays of aperture and dipole antennas located around a cylinder of radius 23.6 cm concentric with the central axis of the model are assumed for the calculations to simulate the APAs that are commonly used. Test runs on the calculation of fields in the water-filled interaction space and with homogeneous circular- and elliptical-cylinder phantoms correlate well with the experimental data (Turner, 1984 a,b). The anatomically-based model of the torso shown in Fig. 13 is then used for SAR calculations. Shown on the left are the layer numbers used for the model and for displaying the calculated SARs. Three cases have been considered for the parallel-plate applicators. In all cases the frequency is 100 MHz and the total incident power is 100W.

In Fig. 14, the layer-averaged SARs are shown for the model of the torso for three cases; namely: (a) 8 radiating elements each with a separation of 33δ or 43.23 cm, (b) 8 elements, each with a separation of 15δ or 19.65 cms, and (c) 5 elements on the side close to the liver to focus the energy preferentially into this organ. The mass densities for the various tissues given in Table I are used to calculate the weights of the individual cells which are then used to convert the volume densities $1/2\ \sigma E^2$ of absorbed power to the SARs for the respective cells. As expected, the energy absorption is somewhat more focused in the central layers for the narrower apertures of separation, 15δ, than for the wider aperture APA. Fig. 15 gives the contours of the calculated SARs for layer #25 (of Fig. 13) corresponding to the centerline of the apertures for the various cases identified as (a), (b) and (c) above. Similar diagrams on the SAR distributions are, of course, available for each of the other layers as well.

In Fig. 16 is the comparison of normalized layer-averaged SARs for the torso model for an 8 aperture APA (case b of Fig. 14) and an 8 dipole APA at a frequency of 100 MHz. Identical dipole length and aperture separations each of 15δ or 19.65 cm have been used for the calculations. Consistent with experimental observations (Turner - personal communication), fairly similar layer-averaged SAR distributions are calculated for aperture- and dipole-type APAs.

As aforementioned, we have recently developed an anatomically-based thermal block model of man (Gandhi and Hoque - to be published) and have used it to calculate the temperature distributions for exposure to plane-wave irradiation conditions. Since the cell sizes of this thermal model are compatible with those used here, SARs obtained here could be used to calculate the temperature distributions for the various APAs. It will also be our attempt in the future to use inhomogeneous models that are obtained from the CT scans of an individual patient so that the size and the location of the tumor could be properly modeled.

CONCLUDING REMARKS

The numerical models for SAR calculations have reached a level of sophistication that would have been unthinkable just a few years ago. These models are consequently beginning to be applied to some very realistic situations both in assessing RF safety and for medical applications. The 3-D thermal models are not as sophisticated at the present time as are the models for SAR calculations. In addition to improvements in the thermal models, future efforts are also needed in using individualized models based on the CT scans of a patient so that size and location of the tumor can be properly modeled.

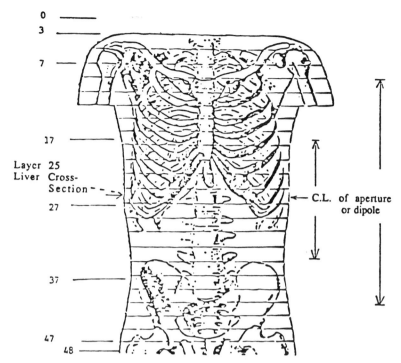

Fig. 13. Diagram of the human torso showing the levels at which data were calculated using the FDTD method. Also shown are the relative sizes of the apertures or dipoles.

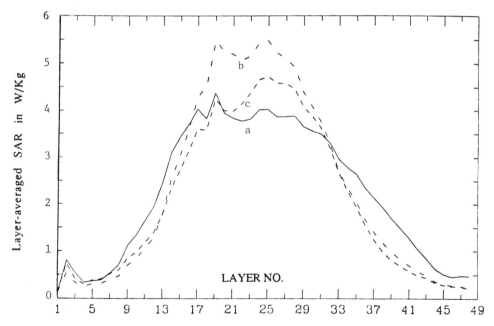

Fig. 14. Layer-averaged SAR distributions for the model of the torso: (a) 8 parallel-plate applicators, each with a separation = 43.23 cm (33δ), (b) 8 parallel-plate applicators each with a separation = 19.65 cm (15δ), (c) 5 elements surrounding the liver are energized, each with a separation = 19.65 cm (15δ). Frequency = 100 MHz. Input power = 100 W.

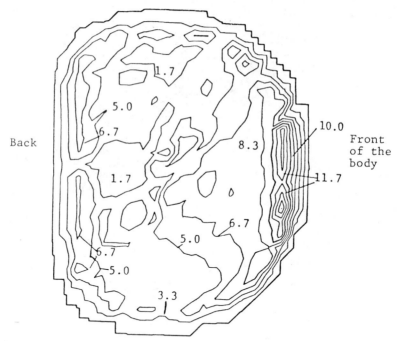

Fig. 15a. SAR distributions for layer 25 corresponding to the central line of the applicators. Eight parallel-plate applicators, each with a separation = 43.23 cm (33δ).

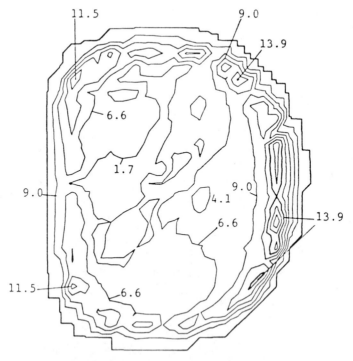

Fig. 15b. SAR distributions for layer 25 corresponding to the central line of the applicators. Eight parallel-plate applicators, each with a separation = 19.65 cm (15δ).

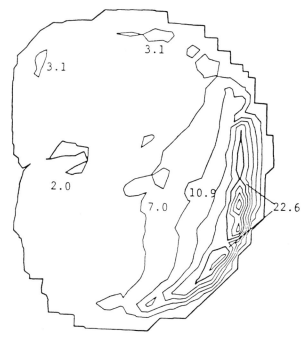

Fig. 15c. SAR distributions for layer 25 corresponding to the central line of the applicators. Five elements surrounding the liver are energized, each with a separation = 19.65 cm (15δ). Frequency = 100 MHz. Input power = 100 W.

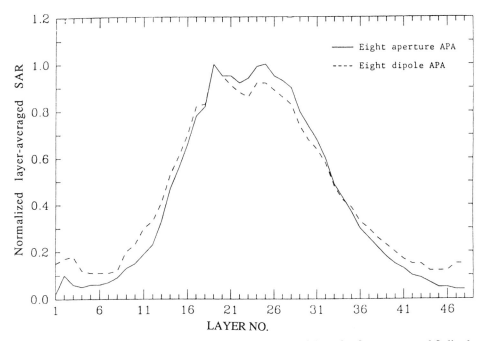

Fig. 16. Comparison of the normalized layer-averaged SARs for 8 aperture and 8 dipole APAs. Frequency = 100 MHz.

REFERENCES

American National Standards Institute (1982). Safety levels with respect to human exposure to radio-frequency electromagnetic fields, 300 kHz to 100 GHz.ANSI Committee C95.1, New York, New York 10017.

Armitage, D. W., Leveen, H. H., and Pethig, R., 1983, Radio-frequency induced hyperthermia: Computer simulation of specific absorption rate distributions using realistic anatomical models. Phy. in Med. and Bio. 28:31.

Borup, D. T., Gandhi, O. P., and Johnson, S. A. (Submitted for publication in IEEE Trans. Microwave Theory & Tech.). A novel sinc basis FFT-CG method for calculating EM power deposition in anatomically-based man models.

Chatterjee, I. and Gandhi, O. P., 1983, An inhomogeneous thermal block model of man for the electromagnetic environment. IEEE Trans. Biomed. Eng. BME-30: 707.

Chen, J. Y. and Gandhi, O. P., 1989a, Electromagnetic deposition in anatomically-based model of man for leakage fields of a parallel-plate dielectric heater. IEEE Trans. Microwave Theory and Tech. MTT-37:174.

Chen, J. Y. and Gandhi, O. P., 1989b, RF currents induced in an anatomically-based model of human for plane-wave exposures (20-100 MHz), Health Phy. 56.

Chen, K. M., and Guru, B. S., 1977, Internal EM field and absorbed power density in human torsos induced by 1-500 MHz EM waves. IEEE Trans. Microwave Theory & Tech MTT-25:746.

DeFord, J. F., Gandhi, O. P. and Hagmann, M. J., 1983, Moment-method solutions and SAR calculations for inhomogeneous models of man with large number of cells. IEEE Trans. Microwave Theory & Tech. MTT-31: 848.

DeFord, J. F.,1985, Impedance method for solving low-frequency dosimetry problems. M.S. Thesis, University of Utah, Salt Lake City, Utah, 84112.

Durney, C. H.,1980, Electromagnetic dosimetry for models of humans and animals: A review of theoretical and numerical techniques. Proc. IEEE 68:33.

Elliott, R. S., Harrison, W. H., and Storm, F. K., 1982, Hyperthermia: Electromagnetic heating of deep-seated tumors. IEEE Trans. Biomed. Eng. BME-29:61.

Eycleshymer, A. C. and Schoemaker, D. M., 1911, "A Cross-Section Anatomy." D. Appleton and Company, New York and London.

Fletcher, R., 1975, Conjugate gradient methods for indefinite systems. In "Numerical Analysis" (G. A. Watson, ed.) Dundee Lecture Note 506, Springer-Verlag.

Gandhi, O. P., Hagmann, M. J. and D'Andrea, J. A., 1979, Part-body and multibody effects on absorption of radio-frequency electromagnetic energy by animals and by models of man. Radio Sci. 14:6(S):15.

Gandhi, O. P., DeFord, J. F. and Kanai, H., 1984, Impedance method for calculation of power deposition patterns in magnetically-induced hyperthermia. IEEE Trans. Biomed. Eng. BME-31: 644.

Gandhi, O. P., Chen, J. Y., and Riazi, A., 1986, Currents induced in a human being for plane-wave exposure conditions 0-50 MHz and for RF sealers. IEEE Trans. Biomed. Eng. BME-33:757.

Gandhi, O. P. and DeFord, J. F., 1988, Calculation of EM power deposition for operator exposure to RF induction heaters. IEEE Trans. Electromag. Comp. EMC-30: 63.

Gandhi, O. P. and Hoque, M. (submitted to IEEE Trans. Biomed. Eng.) An inhomogeneous thermal block model of man for electromagnetic exposure conditions.

Guy, A. W., Chou, C. K., and Neuhaus, B., 1984, Average SAR and SAR distributions in man exposed to 450-MHz radio-frequency radiation. IEEE Trans. Microwave Theory and Tech. MTT-32:752.

Hagmann, M. J., Gandhi, O. P. and Durney, C. H. (1979). Numerical calculation of electromagnetic energy deposition for a realistic model of man. IEEE Trans. Microwave Theory & Tech. MTT-27: 804.

Hagmann, M. J. and Gandhi, O. P., 1979, Numerical calculation of electromagnetic energy deposition in models of man with grounding and reflector effects. Radio Sci. 14:6(S):23.

Harrington, R. F., 1961, "Time Harmonic Electromagnetic Fields" McGraw-Hill Book Company, New York.

Harrington, R. F., 1968, "Field Computations by Moment Methods." The Macmillan Company, New York.

Hestenes, M. and Stiefel, E., 1952, Method of conjugate gradients for solving linear systems. J. Res. Nat. Bur. Standards 49:409.

Hill, D. A., 1984, The effect of frequency and grounding on whole-body absorption of humans in E-polarized radio-frequency fields. Bioelectromagnetics 5:131.

Holland, R., 1977, THREDE: A Free-field EMP coupling and scattering code. IEEE Trans. Nuclear Sci. NS-24:2416.

Kunz, K. S. and Lee K-M., 1978, A three-dimensional finite-difference solution of the external response of an aircraft to a complex transient EM environment: Part 1 - The method and its implementation. IEEE Trans. Electromag. Comp. EMC-20:328.

Livesay, D. E. and Chen, K. M. 1974, Electromagnetic fields induced inside arbitrary-shaped bodies. IEEE Trans. Microwave Theory & Tech. MTT-22:1273.

Manning, M. R. and Gerner, E. W., 1981, Chapter 21, Interstitial thermoradiotherapy. In "Hyperthermia in Cancer Therapy", F. K. Storm, editor.

Mur, G., 1981, Absorbing boundary conditions for the finite-difference approximation of the time-domain electromagnetic field equations. IEEE Trans. Electromag. Comp. EMC-23: 377.

Orcutt, N. E. and Gandhi, O. P., 1988, A 3-D impedance method to calculate power deposition in biological bodies subjected to time-varying magnetic fields. IEEE Trans. Biomed. Eng. BME-35:577.

Orcutt, N. E. and Gandhi, O. P., 1989, Use of the impedance method to calculate 3-D power deposition patterns for hyperthermia with capacitive plate electrodes, IEEE Trans. Biomed. Eng. BME-36.

Sarker, T. K., Siarkiewicz, K. R., and Stratton, R. F., 1981, Survey of numerical methods for solution of large systems of linear equations for electromagnetic field problems. IEEE Trans. Antennas & Prop. AP-29:847.

Spiegel, R. J., Deffenbaugh, D. M. and Mann, J. E., 1980, A thermal model of the human body exposed to an electromagnetic field. Bioelectromagnetics 1: 253.

Spiegel, R. J., 1984, A review of numerical models for predicting the energy deposition and resultant thermal response of humans exposed to electromagnetic fields. IEEE Trans. on Microwave Theory & Tech. MTT-32:730.

Spiegel, R. J., Fatmi, M. B. E. and Kunz, K. S., 1985, Application of a finite-difference technique to the human radio-frequency dosimetry problem. J. Microwave Power 20: 241.

Stolwijk, J. A. J., 1970, Mathematical model of thermoregulation. In "Physiological and Behavioral Temperature Regulation." (J. D. Hardy, A. P. Gagge, and J. A. J. Stolwijk, eds.) Charles C. Thomas, Springfield, Illinois.

Stolwijk, J. A. J. and Hardy, J. D., 1977, Control of body temperature. In "Handbook of Physiology." Section 9: Reactions to Environmental Agents. (D. H. K. Lee, ed.) American Physiological Society, Bethesda, Maryland.

Stolwijk, J. A. J., 1980, Mathematical models of thermal regulation. Ann. N. Y. Acad. Sci. 335: 98.

Sullivan, D. M., Borup, D. T. and Gandhi, O. P., 1987, Use of the finite-difference time-domain method in calculating EM absorption in human tissues. IEEE Trans. Biomed. Eng. BME-34:148.

Sullivan, D. M., Gandhi, O. P. and Taflove, A., 1988, Use of the finite-difference time-domain method for calculating EM absorption in man models. IEEE Trans. Biomed. Eng. BME-35:179.

Taflove, A. and Brodwin, M. E., 1975a, Numerical solution of steady-state electromagnetic scattering problems using the time-dependent Maxwell's equations. IEEE Trans. Microwave Theory & Tech. MTT-23: 623.

Taflove, A. and Brodwin, M. E., 1975b, Computation of the electromagnetic fields and induced temperatures within a model of the microwave irradiated human eye. IEEE Trans. Microwave Theory & Tech. MTT-23: 888.

Taflove, A., 1980, Application of the finite-difference time-domain method to sinusoidal steady-state electromagnetic-penetration problems. IEEE Trans. Electromag. Comp. EMC-22:191.

Turner, P. F., 1984a, Regional hyperthermia with an annular phased array. IEEE Trans. Biomed. Eng. BME-31:106.

Turner, P. F., 1984b, Hyperthermia and inhomogeneous tissue effects using an annular phased array. IEEE Trans. Microwave Theory & Tech. MTT-32: 874.

Umashankar, K. and Taflove, A., 1982, A novel method to analyze electromagnetic scattering of complex objects. IEEE Trans. Electromag. Comp. EMC-24: 397.

Wang, C. Q. and Gandhi, O. P., 1989, Numerical simulation of annular-phased arrays for anatomically-based models using the FDTD method. IEEE Trans. Microwave Theory & Tech. MTT-37:118.

Yee, K. S., 1966, Numerical solution of initial boundary value problems involving Maxwell's equations in isotropic media. IEEE Trans. Antennas & Prop. AP-14:302.

Zhu, X. L. and Gandhi, O. P., 1988, Design of RF needle applicators for optimum SAR distributions in irregularly-shaped tumors. IEEE Trans. Biomed. Eng. BME-35:382.

CONTRIBUTORS

Bernardi Paolo
Department of Electronics
University of Rome "La Sapienza"
Via Eudossiana, 18
00184 Rome, Italy

Cleary F. Stephen
Department of Physiology
Medical College of Virginia
Virginia Commonwealth University
P.O. Box 55IMCV Station
23298 Richmond, Virginia, USA

d'Ambrosio Guglielmo
Department of Electronic Engineering
University of Naples
Via Claudio 21
80125 Naples, Italy

D'Inzeo Guglielmo
Department of Electronics
University of Rome "La Sapienza"
Via Eudossiana, 18
00184 Rome, Italy

Franceschetti Giorgio
IRECE-CNR
Via Claudio 21
80125 Naples, Italy

Gandhi P. Om
Department of Electrical Engineering
University of Utah
3280 Merrill Engineering Building
Salt Lake City, Utah 84112, USA

Gil Jerzy
MMA Postgraduate Medical School
128 Szaserow
00 909 Warsaw, Poland

Grandolfo Martino
National Institute of Health
Physics Laboratory
Viale Regina Elena, 299
00161 Rome, Italy

Kanda Motohisa
National Institute of Standards
and Technology
Electromagentic Fields Division
Boulder, Co 80303, USA

Mild Kjell Hannson
National Institute of Occupational Health
Box 6104
S-900 06 Umea, Sweden

Obara Tadeusz
Department of Biological Effects
of Non Ionizing Radiation
Center for Radiobiology and Radiation Safety
128 Szaserow
00 909 Warsaw, Poland

Szmigielski Stanislaw
Department of Biological Effects
of Non Ionizing Radiation
Center for Radiobiology and Radiation Safety
128 Szaserow
09 Warsaw, Poland

Tofani Santi
USSL 40 Laboratorio di Sanita' Pubblica
Sezione di Fisica
Via Lago San Michele 11
10015 Ivrea, Italy

Vecchia Paolo
National Institute of Health
Physics Laboratory
Viale Regina Elena, 299
00161 Rome, Italy

INDEX

Absorption parameters, 8
  absorbed energy (dose), 9
  average absorbed power, 10
  average absorbed power mass density, 9
  average absorbed power volume density, 9
  local absorbed power mass density, 9
  local absorbed power volume density, 9
  specific absorption rate, 9
Adaptation, 150
Adaptation mechanisms, 84
Adaptive reactions, 83
Adenosine triphosphate, 71
ALARA principle, 121, 127
American Conference of Governmental Industrial Hygienists, 121
American National Standards Institute, 12, 120
Anechoic chamber
  typical errors, 186
Anechoic material, 178
ANSI, see American National Standards Institute
Antenna gain, 181
Antigen-antibody complexes, 68
Antineoplastic systems, 82
Aplysia pacemaker cells, 69
Atomic polarization, 28, 31
Austrian Standards Institute, 104

B-lymphocytes, 68
Bioheat equation, 193
Biological dosimeter, 136
Biological effects, 59 (see also Interaction mechanisms)
  allogenic cytotoxicity, 68
  ATPase, 65
  cell growth, 71
  compensative reactions, 150
  decarboxilase, 92
  delayed effects, 136
  electrophoretic mobility, 67

Biological effects (continued)
  erythrocytes, 64
  erythrocytes membranes, 65
  genetically determined cancer, 82
  genomic effects, 73
  hemolysis, 66
  histone protein kinase, 92
  hot spots, 61
  internal viscosity, 67
  leukocytes, 68
  long-term effects, 81, 136
  mammalian cells, 63
  membrane fluorescent probes, 67
  neoplastic diseases, 82
  nonspecific thermal effects, 60
  nucleic acid, 92
  nucleic acid turnover, 92
  phatological changes, 150
  physiological adaptation, 150
  plasma membrane, 73
  potassium permeability, 64
  protein turnover, 92
  protein-lipid interactions, 66
  red cell membranes, 67
  reparative regeneration, 150
  rf-specific thermal effects, 60
  salivary gland cells, 73
  serum electrolytes, 65
  sodium permeability, 64
  T-limphocytes, 68
  transcriptional patterns, 73
  uncertain tissues, 138
Biphasic windowed effects, 72
Block model
  human body, 194
Brain energy metabolism, 70
Brownian motion, 34

Calcium ions flux, 47, 72
  cat cerebral tissue, 72
  chick cerebral tissue, 72
Calibration methods
  accuracy, 185
  free space standard, 178
  guided wave, 182

Calibration methods (continued)
  reverberating chamber, 184
  transfer standard, 184
Canadian Bureau of Radiation and Medical Devices, 106
Cancer morbidity
  hemato-immunologic neoplasms, 85
  hemopoietic neoplasms, 90
  leukaemia, 84
  lymphatic neoplasms, 90
  neurologic neoplasms, 85
  pancreatic cancer, 85
  retrospective analyses, 85
Carcinogenicity, 71, 81
  benzopyrene, 92
  cancer morbidity, 81
  di-ethyl-nitrosoamine, 94
  environmental factors, 82, 84
  experimental investigations, 92, 93
  methilcholantrene, 94
  murine embryonic fibroblasts, 92
  nonthermal fields, 92
  occupational factors, 82
  phorbol esters, 92
  tumor-promoting activity, 94
  x-rays, 92
Catalytic activity, 71
CDRH, see Center for Devices and Radiological Health
Cell membrane, 40
Center for Devices and Radiological Health, USA, 106
Center for Radiobiology and Radiation Safety, Poland, 81, 135
Central Research Institute of the Electric Power Industry, Japan, 161
Chara braunii, 70
Chromosomal damage, 69
CIGRE, see International Conference on Large High Voltage Electric Systems
Clausius-Mossotti-Lorentz theory, 32
Comecon, see Council for Mutual Economic Assistance
Compensatory mechanisms, 84
Compensatory reactions, 83
Complex relative permittivity, 29
Contact hazards, 15
  finger contact, 15
  grasping contact, 15
  VLF to HF frequency band, 15
Council for Mutual Economic Assistance, 197
Cyclotron resonance, 48, 72
Cyclotron resonance frequency, 48

Data logging, 178

DC electric fields, 73
  acetylcholine receptors, 73
  dielectric breakdown, 74
  electrophoretic movement, 73
  embryonic neurites, 73
  glycoproteins, 73
  membrane permeability, 74
  osmotic lysis, 74
  receptor aggregates, 73
  xenopus myoblasts, 73
Debye (unit), 31
Debye relaxation, 29
  dried haemoglobin, 38
  friction constant, 34
  peptide unit, 36
  relaxation frequency, 34
  relaxation time, 34
  water, 33
Densitometry, 61
Dielectric Polarization, 28
Dipolar structure
  protein molecule, 33
  water, 33
Dipole moment, 28
DNA synthesis, 71
Dosimetry, 11, 61, 101, 193
  biconjugate gradient method, 195
  conjugate gradient method, 193
  fast Fourier transform, 193
  finite-difference time-domain method, 193, 196
  grounded conditions, 203
  hot-spot range, 101
  impedance method, 194, 203
  method of moments, 193, 194
  numerical methods, 193
  partial body resonances, 101
  power transmission lines, 203
  prolate spheroid models, 101
  resonance range, 101
  sub-resonance range, 101
  surface absorption range, 101
  temperature distributions, 193
  thermal block models, 193
  ungrounded conditions, 203
  whole body resonances, 101

Electric networks
  operation voltages, 156
Electrical appliances, 153
Electromagnetic compatibility
  reverberating chambers, 178
Electromagnetic fields
  basic definitions, 1
Electromagnetic quantities
  angular frequency, 3
  circular polarization, 7
  conductivity, 2
  current density, 2
  dielectric losses, 4

Electromagnetic quantities (continued)
  electric field, 1
  electric induction, 2
  frequency, 3
  magnetic field, 2
  magnetic induction, 1
  ohmic losses, 4
  permeability, 2
  permittivity, 2
  power, 4
  power flux, 6
  power surface density, 6
  time period, 3
  wavelength, 4
Electromagnetic spectrum, 5, 140
  most important applications, 5
Electron transport proteins, 71
Electronic polarizability, 31
Electrophoretic mobility, 67
ELF, see Extremely low frequency fields
Emission standard, 103
Environmental Health Criteria Documents
  extremely low frequency fields, 127
  magnetic fields, 171
  radiofrequency and microwave, 127
Environmental Health Criteria Programme, 126
Environmental Protection Agency, USA, 123
EPA, see Environmetal Protection Agency
Epidemiological studies, 81
  acute myelocytic leukaemia, 88
  blood-forming system, 88
  brain tumors, 88
  chronic lymphocytic leukaemia, 88
  chronic myelocytic leukaemia, 88
  colorectal neoplasms, 88
  intestinal neoplasms, 88
  leukaemias, 88
  lung cancer, 88
  lymphatic system, 88
  lymphomas, 88
  lymphosarcomas, 88
  malignant lymphogranulomatosis, 88
  melanoma, 88
  plasmocytomas, 88
  prospective analyses, 88
  retrospective analyses, 88
  skin neoplasms,
  stomach, 88
Ergonomy, 142
Exposure evaluation, 175

Exposure limits, see Protection guides
Exposure methods, 175, 178
  bidirectional coupler, 180
  bolometric sensors, 180
  directive antenna, 180
  function generator, 180
  guided wave, 182
  open-ended waveguides, 181
  power meter, 180
  pyramidal horns, 181
  rectangular-waveguide transmission cells, 182
  reverberating chamber, 184
  TEM cells, 180
Exposure parameters, 8
  amplitude modulation characterstics, 137
  cancer related aspects, 84
  polarized plane-wave, 13
Exposure standard, 103
Exposure standards, see Protection guides
Exposure zones, 144, 148
Extremely low frequency fields, 153
  amplitude windows, 72
  collagen synthesis, 73
  fibroblast proliferation, 73
  fibroplasia in vitro, 73
  frequency windows, 72
  physiological consequences, 72
  windowed responses, 72

Faraday's law, 200
FDA, see Food and Drug Administration
Fertilization, 71
Fiber optics, 177
Field strength meters
  antenna, 176
  detector, 176
  dynamic range, 177
  electro-optic modulator, 176
  filter network, 176
  frequency response, 177
  isotropic probe, 176
  metering unit, 176
  non-linear circuitry, 176
  principal characteristics, 177
  semiconductor diodes, 176
  short dipole, 176
  small loop, 176
  sources of errors, 187
  thermocouples, 176
Finnish Centre for Radiation and Nuclear Safety, 111
Firing rate of neurons, 69
Food and Drug Administration, USA, 125
Free energy, 43

Free-space impedance, 184
Frequency domain, 30

German Electrotechnical Commission on DIN and VDE, 172
Giant algal cells, 70
Guide numbers, 107
Guidelines, see Protection guides

Hair motion, 109
Health and Safety Executive, UK, 118
Health and Welfare Canada, 132
Helix pomatia, 70
High risk groups of people, 142
High voltage transmission lines, 153
Homeostasis, 83
Hot spots, 60
HSE, see Health and Safety Executive
Human glioma cells, 71
HWC, see Health and Welfare Canada
Hydrogen bond, 35
  in water, 35
Hyperpolarization, 70
Hypersensitivity, 90, 141
  in the vicinity of power lines, 138
  in the vicinity of Radio/TV transmitters, 138
Hyperthermia
  annular phased arrays, 206
  capacitor-type electrodes, 203
  dipole antennas, 206
  needle applicators, 203

IEC, see International Electrotechnical Commission
Induced currents
  bare feet, 15
  grounded man, 204, 205
  isolated man, 204, 205
  local current densities, 205
  standing human being, 13
  threshold values, 15
  touching vehicles, 13
  wearing leather-soled shoes, 15
  wearing rubber-soled shoes, 15
INFN, see Istituto Nazionale di Fisica Nucleare, Italy
INFN-Sezione di Torino, Italy, 175
INFN-Sezione Sanita', Italy, 99, 153
INIRC, see International Non Ionizing Radiation Committee
INRS, see Institut National de Recherche et de Securite'
Institut National de Recherche et de Securite', France, 112

Institute of Labour Medicine, Poland, 149
Instrumentation
  accuracy, 185
  field strength meters, 176
  near-field measurements, 176
  sources of errors, 187
Interaction mechanisms, 50
  albumin, 37
  alpha lipoproteins, 37
  amphora coffeaformis, 49
  beta lipoproteins, 37
  biopolymers, 37
  camphor-binding protein, 44
  cell cycle, 71
  channel proteins, 44
  chick myotube, 45
  cholinergic activity, 44
  diatoms, 49
  dried proteins, 37
  DNA, 36
  enzymatic reactions, 50, 51
  erythrocytes, 64
  globulin, 37
  gramicidine A receptors, 44
  hemoglobin, 37
  hydroxylic compounds, 33
  ions, 47
  leukocytes, 68
  ligands, 47
  macroscopic level, 28
  membrane surfaces, 50
  microscopic level, 27, 31
  neural cells, 69
  olfactory epithelium, 44
  phosphocholine, 42
  phospholipids, 41
  squid axon membrane, 51
International intercomparisons, 186
International Conference on Large High Voltage Electric Systems, 155
International Electrotechnical Commission, 126
International Non Ionizing Radiation Committee, 12
International Programme on Chemical Safety, 126, 170
International Radiation Protection Association, 12
International Union of Producers and Distributors of Electric Energy, 154
Interspecies extrapolation, 60
IRPA, see International Radiation Protection Association
Istituto Nazionale di Fisica Nucleare, 187

Laboratorio di Sanita' Pubblica,

Ivrea-Italy, 175, 187
Lawrence Livermore National Laboratory, 121
Leakage fields
  parallel-plate dielectric heater, 205
Long-Term Programme on Non-Ionizing Radiation Protection, 127

Magnetic resonance imaging, 200
Mammalian neurons, 69
Man model
  cross-sectional diagrams, 198
Man-made EM fields, 81
  environmental exposure, 139
Matrix equation
  conjugate gradient method, 194
  image theory, 194
  Gauss elimination, 194
Maxwell's equations, 193, 196
Mean squared field strength, 177
Medical examinations, 136, 138
    (see also Medical Surveillance)
Medical surveillance, 104, 112, 115, 138, 148
  behavioral profile, 122
  emotional profile, 122
  evaluation of cardiovascular function, 122
  evaluation of neurologic function, 122
  evaluation of the eyes, 122
  evaluation of the skin, 122
  laboratory examinations, 122
  medical hystory, 122
  preplacement examinations, 122
  work hystory, 122
Membrane dielectric properties, 70
Membrane fluorescent probes, 67
Membrane resting potentials, 70
Microshocks, 109
Microwave ovens, 117, 125
Ministry of Health of the USSR, 167, 169
Ministry of International Trade and Industry, Japan, 160
Mitochondrial enzymes, 71
Mitogenesis, 69
MMA Postgraduate Medical School, Poland, 81
Molecular polarizability, 31

National Board of Occupational Safety and Health, Finland, 111
National Board of Occupational Safety and Health, Sweden, 116
National Council on Radiation Protection, USA, 12
National Electric Safety Code, USA, 24, 165
National Institute for Occupational Safety and Health, USA, 122
National Institute of Health, Physics Laboratory, Italy, 99, 153
National Institute of Occupational Health, Sweden, 99
National Institute of Radiation Protection, Sweden, 117
National Institute of Standards and Technology, USA, 175, 181
National Radiological Protection Board, UK, 117, 162
Natural background, 81
Natural Resource and Environment Committee, Victoria-Australia, 158
NBOSH, see National Board of Ocpational Safety and Health
NCRP, see National Council of Radiation Protection
Neoplasms, 81
  children, 82
  cocarcinogenesis, 82
  environmental exposure, 91
  exposure to magnetic fields, 82
  high-voltage power lines, 82
  immunosuppressive agents, 82
  initiation, 82
  occupational exposure, 86, 87
  oncogens, 82
  phorbol esters, 82
  promotion, 82
  stressors, 82
  tumors, 82
NESC, see National Electric Safety Code
Neuro-endocrine-lymphatic network, 83
Neutrophil viability, 69
Nicotinamide adenine dinucleotide, 70
NIOSH, see National Institute for Occupational Safety and Health
NIST, see National Institute of Standards and Technology
Nitella flexilis, 70
Non-specific stress reaction, 84
Non-specific symptoms, 138
Non-stationary fields (Polish standard), 114, 143, 144
Nonthermal effects, 63, 73
  low frequency electric fields, 73
  low frequency magnetic fields, 73

NRPB, see National Radiological Protection Board

Occupational Safety and Health Agency, USA, 123
Oesterreichisches Normungsinstitut, 133
Open-circuit voltages, 13
Oscillation frequency, 37
OSHA, see Occupational Safety and Health Agency

Pacemaker wearers, 109, 116
Patch-clamp technique, 44
Perception of electric fields, 109
Perception of magnetic fields, 109
Performance standard, 103
Phagocytic activity, 69
Physical surveillance
  near field evaluation errors, 175
Power density, 180
  equivalent plane wave, 177
Product emission standards, 125
Proliferative capacity, 69
Protecting devices, 142
Protection guides (RF and MW)
  Australia, 103
  Austria, 104
  Belgium, 105
  Canada, 105
  Comecon, 135, 143
  Comecon (proposed), 147
  Czechoslovakia, 107
  European Community, 126
  Federal Republic of Germany, 109
  Finland, 111
  France, 112
  IRPA, 126
  Italy, 112
  Norway, 113
  Piemonte region (Italy), 113
  Poland, 114, 135, 143
  Poland (proposed), 147
  rationale for the eastern European countries, 135
  Sweden, 116
  United Kingdom, 117
  USA, 119
  USSR, 124, 135, 140
  WHO, 126
Protection guides (50/60 Hz)
  Australia, 156
  Comecon (proposed), 147
  Czecholosvakia, 158
  Federal Republic of Germany, 158
  IRPA, 170
  Japan, 160
  Poland, 161
  Poland (proposed), 147
  United Kingdom, 162

Protection guides (50/60 Hz) (continued)
  USA, 165
  USSR, 166

Quasi-static approximation, 193, 199
  impedance method, 199

Radiative far field region, 7
Radiative field region, 7
Radiative near field region, 7
Radiators
  directive antennas, 178
  horns, 178
  open-ended guides, 178
Radiofrequency burns, 15
Radiofrequency shocks, 15
Reactive field region, 7
Reflection coefficient, 6, 179
Reflectivity, see Reflection coefficient
Relaxation time, 28
Right-of-way (RoW), 157, 158, 160, 165
Risk-benefit analysis, 154
RNA synthesis, 71
Root mean square value, 3, 176
  rms phasor, 3
Root sum of square, 176
Rotating antenna, see Non-stationary fields

Safety coefficients, 139
Safety shoes, 206
Sanitary protection zone, 168
SAR, see Specific absorption rate
Scaling concepts, 13, 15
Spark discharges, 109
Specific absorption rate, 101, 102, 193
  in the ankle, 13
  below resonance, 11
  effective physical areas, 13
  grounded conditions, 12
  induction heater, 200
  in the leg, 13
  resonant region, 11
  scaling concepts, 13
  surface temperature elevation, 15, 17
  thermal implications, 15
  ungrounded conditions, 12
  variations with frequency, 102
  in the wrist, 15, 18
Spermatozoa, 71
Stability condition, 197
Standards, see Protection guides
Standard fields, 178
Standards Association of Australia, 133

State Institute of Radiation Hygiene, Norway, 113
Stationary antenna, see Stationary fields
Stationary fields (Polish standard), 114, 143, 144
Stokes' law, 34
Stratton-Chu formula, 181
Surface effects, 109

TEM cells
  typical errors, 186
Thermal model
  of the human body, 193
Threshold Limit Values, 121
Time domain, 30
Time-harmonic fields
  angular frequency, 3
  frequency, 3
  phasor quantities, 3
  time period, 3
Time-harmonic plane waves, 4
  attenuation constant, 4
  free space propagation constant, 6
  incident wave, 6
  intrinsic impedance, 6
  phase constant, 4
  phase velocity, 4
  propagation constant, 4
  reflected wave, 6
  transmitted wave, 6
  wavelength, 4
Time window response, 68
Tissue properties
  blood, 199
  brain, 199
  cartilage, 199
  eye, 199
  fat bones, 199
  heart, 199
  intestine, 199
  kidney, 199
  liver, 199
  lung, 199
  muscle, 199
  nerve, 199

Tissue properties (continued)
  pancreas, 199
  skin, 199
  spleen, 199
TLVs, see Threshold Limit Values
Transcriptional activity, 73
Transition probability, 37
Transport activity, 71
Tumor promotion, 71

UNEP, see United Nations Environment Programme
UNIPEDE, see International Union of Producers and Distributors of Electrical Energy
United Nations Conference on the Human Environment, 126, 170
United Nations Environment Programme, 126, 170
University of Naples, Italy, 1
University of Rome "La Sapienza", Italy, 27
University of Utah, USA, 11, 193

VDTs, see Video display terminals
Video display terminals, 138
Virginia Commonwealth University, USA, 59

WHO, see World Health Organization
WHO Environmental Health Criteria Programme, 170
WHO Regional Office for Europe, 127
World Health Organization, 126
Worst case method, 185

Yaghjian correction term, 182
Yee lattice, 196

Zone boundaries (Polish standard)
  danger zone, 114, 148
  hazardous zone, 114, 148
  intermediate zone, 114, 148
  safe zone, 114, 148